2001
THE LOST SCIENCE

The science and technology of the most influential film ever made.

WRITTEN BY

ADAM K. JOHNSON

with additional text by

Frederick I. Ordway III

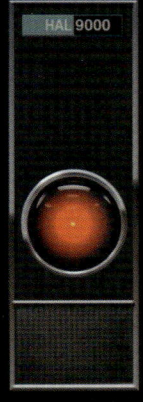

CONTENTS

INTRODUCTION
3

PREFACE
4

SATELLITES
5

ORION III SPACEPLANE
15

TITOV SPACEPLANE/SHUTTLE
21

SPACE STATION V
22

ARIES IB SHUTTLE
25

CLAVIUS BASE
32

EXPLORING THE MOON
34

TYCHO MAGNETIC ANOMALY 1
39

THE DISCOVERY
41

- AE-35
- GASEOUS FISSION PLASMA REACTOR
- COMMAND MODULE
- THE CENTRIFUGE
- THE PODBAY
- PODBAY TEST BENCH
- EVA PODS
- POD INTERIOR
- THE ATHENA ROOM
- EQUIPMENT STORAGE ROOM
- DISCOVERY EVA SUIT
- ENVIRONMENTAL CONTROL UNIT
- AIRLOCK
- HAL 9000

THE BEST MODEL SHOP IN HISTORY
87

APPENDIX 1
96

- PRODUCTION PHOTOS
- POPULAR MECHANICS

APPENDIX II
102

- TECHNICAL ADVISORS

APPENDIX III
106

- HONEYWELL PROSPECTUS
- REFERENCES
- ACKNOWLEDGEMENTS

INTRODUCTION

In January 1965, while staying at the Harvard Club in Manhattan, Frederick I. Ordway III arranged a social meeting with his 'old friend' Arthur C. Clarke. The next day, at Clarke's insistence, Ordway, and his associate Harry Lange, met with famed film director Stanley Kubrick who quickly invited Ordway to be the Senior Science advisor on his proposed new science fiction epic provisionally entitled '*Journey to the Stars*'. Within two days, the project that would eventually evolve into the film '*2001: A Space Odyssey*' emerged from the meeting between the four men. In the following 2½ years spent working on the film, Stanley and Frederick worked painstakingly on a daily basis to ensure scientific realism. Working at the zenith of the 'space race' their efforts drew upon the most current space travel 'hard science' available. Between January and August 1965, working from Stanley's New York Office "Hawk Films" (also known to the contractors as *Polaris Industries*), Fred contacted a multitude of companies known to be on the forefront of aerospace technology and asked for their assistance on the film. By Early 1966, over one hundred companies had submitted engineering and design proposals to aid the vision of Kubrick and Clarke.

This book is intended as a companion to viewing the film *2001: A Space Odyssey*, as well as a resource for modern day aerospace engineers interested in how the proposed 21st century technologies might have worked if so implemented.

Creating the designs for the film involved over 2400 detailed schematics that were drawn 4 x 6 feet or 3 x 5 feet. These drawings were saved as blue-line prints using ammonia based blue ink that replaced the original pencil or ink lines. Over time, this blue ink fades, and if exposed to light, completely vanishes. The U.S. SPACE & ROCKET CENTER archives still possess about 100 of the known existing 200 blue-line prints (in various states of condition) created for *2001*.

Restoration of the blue-line prints involved creating a digital replica via a precision CCD scanning camera to ensure a distortion-free two-dimensional representation. Then, the process of restoration required removing the artifacts from the image that were *not* actual lines, lettering or other hand drawn elements. This process was very time-consuming! In some cases, due to light exposure fading the blue ink, the naked eye could not see the original lines so an ultra-violet light was used to detect where the hand-drawn elements use to be. In these few cases, the lines were redrawn digitally.

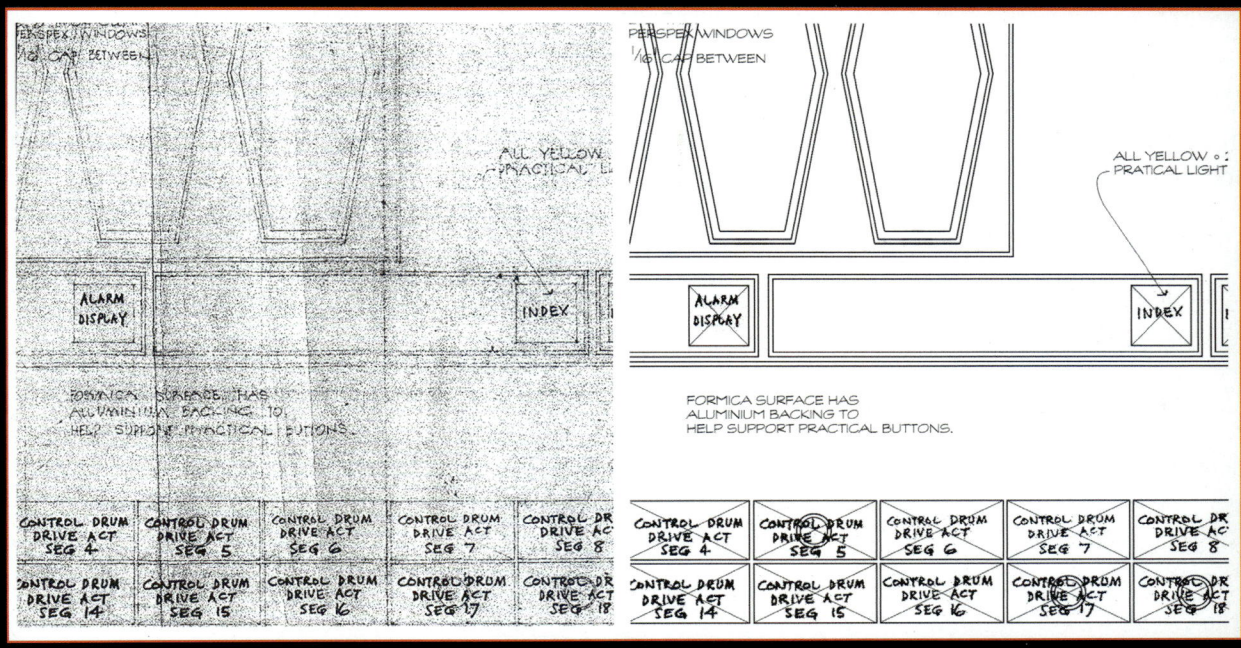

Photographic images in the collection are in excellent condition considering their age. Almost 50 of the images are 11 x 14 or 16 x 20 inches. These photos were taken with a large format *plate camera* as lighting tests for later review by Stanley Kubrick, and are 30 to 50 times the resolution of 35mm film. All images in this book were scanned at 1000 dpi or higher and cannot actually duplicate the true resolution of the original image. The spacecraft images are reproduced here at about 50% loss. All other photographic scans have negligible loss. About 35% of the images were digitally restored via color correction, removal of rips and tears or creases, or enhancement with highly intuitive specialized photo restoration software. The total collection is estimated at 2500 images, and 1500 documents encompassing over 8,000 pages.

The lighting tests during production of *2001* were conducted using multiple exposures of the subject with a neutral jet-black background. Most of the photographs seen here have many different versions – even though they *appear* to be the same photo. For example, the Orion III space-plane was photographed up to twelve times. The changes made between photos used different F-stops (size of the camera aperture or iris). So, the first shot was set at a stop of f-8, second at f-12, third at f-16, and so on. Stanley Kubrick would then review the processed exposures and choose the one that best suited his artistic sensibilities.

A significant portion of the background text presented in the following pages was written by Fred Ordway at the time of the production. Other details are derived from latter-day commentary by Sir Arthur C. Clarke (such details are noted in the text.)

It is my hope that the reader—after absorbing this book—will enjoy viewing, or re-viewing *2001: A Space Odyssey,* and perhaps see the work of these highly talented individuals that created the science and technology for Stanley Kubrick's film in a new light.

Adam K. Johnson

PREFACE

Many people were essential to the creation of the *science* of 2001: A Space Odyssey. It is not possible to list all of these contributors here. The principle people responsible for creating the scientific reality in *2001* are as follows:

Stanley Kubrick:

Born in New York City in 1928, he collaborated in 1964 with Arthur C. Clarke to create a film that would accurately portray space travel based on current scientific knowledge. He rapidly set out to hire the best scientific minds in the world to work on *2001*. Naturally he was led to NASA, and the various companies that contracted for them. Kubrick demonstrated a natural ability to grasp and apply complex scientific theorem and was reading the subject matter at a voracious pace.

Kubrick's obsessive learning ability, as well as his ability to apply what he had learned, earned him great respect from his scientific colleagues, and he was seen as an 'intellectual match'. He oversaw every detail in the creation of the science and technology. His extreme attention to detail is what made *2001* an enormous success and earned the title "most influential film ever made". He died in 1999.

Frederick I. Ordway III:

Born in New York City in 1927, he was the Lead Science Advisor on *2001*. Educated at Harvard University, he did graduate work at the University of Paris, and was awarded an honorary Doctorate of Science from the University of Alabama in Huntsville.

Ordway's interests led him to join the American Rocket Society in 1941 and, a decade later, he joined Wernher von Braun at the Army Ballistic Missile Agency in a liaison capacity with NASA and Department of Defense projects. He lived out his dream being an integral part of the team that launched the Explorer 1 satellite, and he worked with NASA all the way through the Apollo moon launches. Fred continues to work today as a special consultant to NASA.

Harry Lange:

Born in East Germany in 1930, he was the primary Science Illustrator and Concept artist on *2001*. He studied art in West Germany and moved to the U.S. in 1951 where he illustrated flight manuals for the military. Eventually he worked at the Army Ballistic Missile Agency with Ordway and Von Braun on the U.S. space program as a Spacecraft Designer in the 'Future Projects Division'.

After *2001*, Lange continued working in the film business on such projects as *Star Wars: The Empire Strikes Back*, *Star Wars: Return of the Jedi*, *Moonraker*, and *Superman II*. He died in 2008.

Arthur C. Clarke:

Born in England in 1917, he was the novel's writer and co-writer of the screenplay. He served as Adjunct Science Advisor on *2001*. Clarke spent his early years as an auditor at the Board of Education. In WWII, he became a radar specialist and Flight officer. After the war, he earned degrees in mathematics and physics at Kings College in London.

Clarke became Chairman of the *British Interplanetary Society* in 1951 and proposed the idea of permanent satellites in orbit around Earth performing various functions, including communications, telemetry, and weaponry. Clarke wrote award-winning science fiction stories throughout his life, and was keen about working with Kubrick on *2001* from the beginning. Clarke commented that working with Kubrick was 'one of the greatest collaborations of my life'. He died in 2008.

Anthony (Tony) Masters:

Born in England in 1919, he was the Production Designer of *2001*. His career spanned 1949 to 1987. He was Art Director of David Lean's *Lawrence of Arabia* when he was caught by Kubrick's keen eye. He oversaw every detail of the visual aspects of *2001*. Tony was respected by his peers for a broad and immense artistic ability. He went on to design such films as *Papillon* 1973, and *Dune* 1984. He died in 1990.

Stanley Kubrick

Harry Lange, Arthur C. Clarke and Frederick I. Ordway III

Anthony Masters

THE SATELLITES

The film's first space scene shows various "bombs" gracefully orbiting around the earth. They illustrate the global tension between nations, and they accurately predict the current global stalemate due to a fear of terrorism, and the struggle for power between nations. Each satellite represents a different nation. These simple designs were drawn by NASA concept artist Harry Lange. No schematics were ever drawn, and the space vehicles were produced directly from the sketches. The original drawings were white pastel over black parchment paper. The surface of this parchment was deemed too susceptible to handling damage and so the drawings were photographed with a plate camera for departmental distribution. The fine white lines with dimensions (for the model department) were drawn directly onto the photograph with a wax pencil.

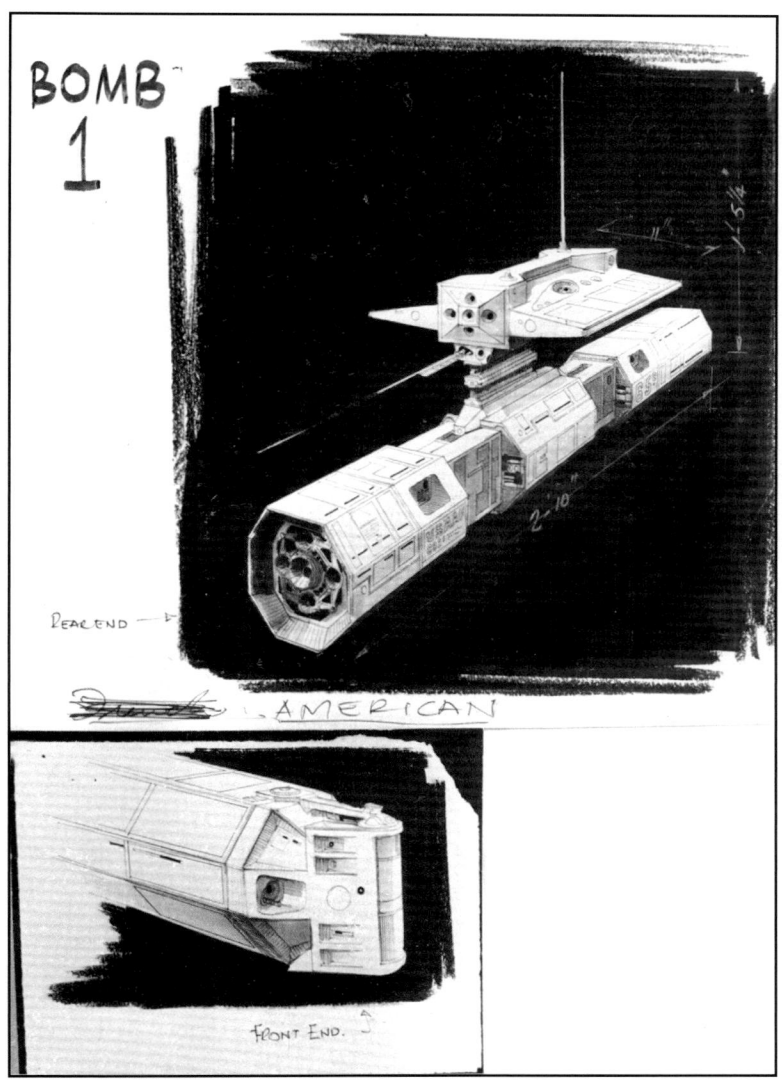

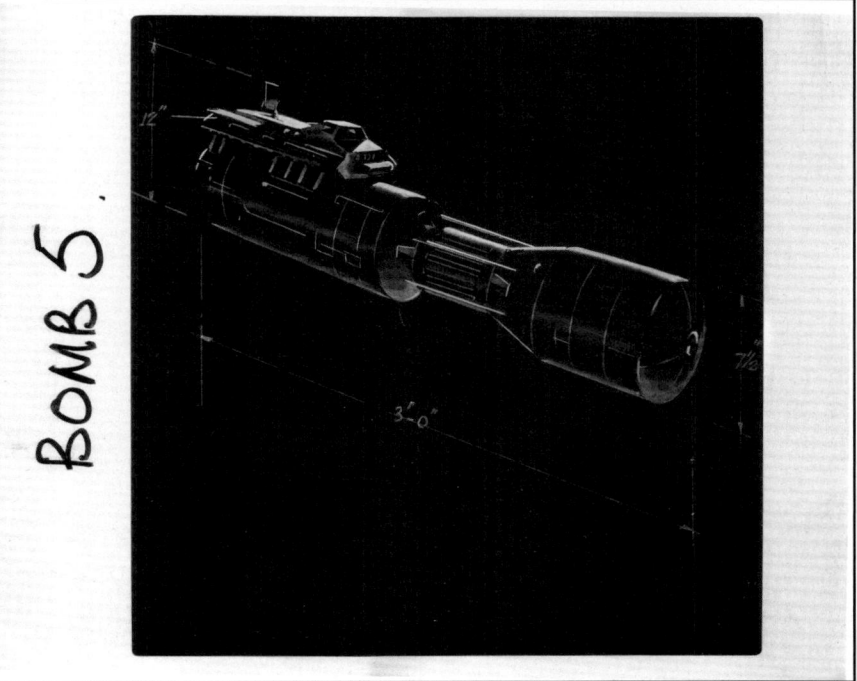

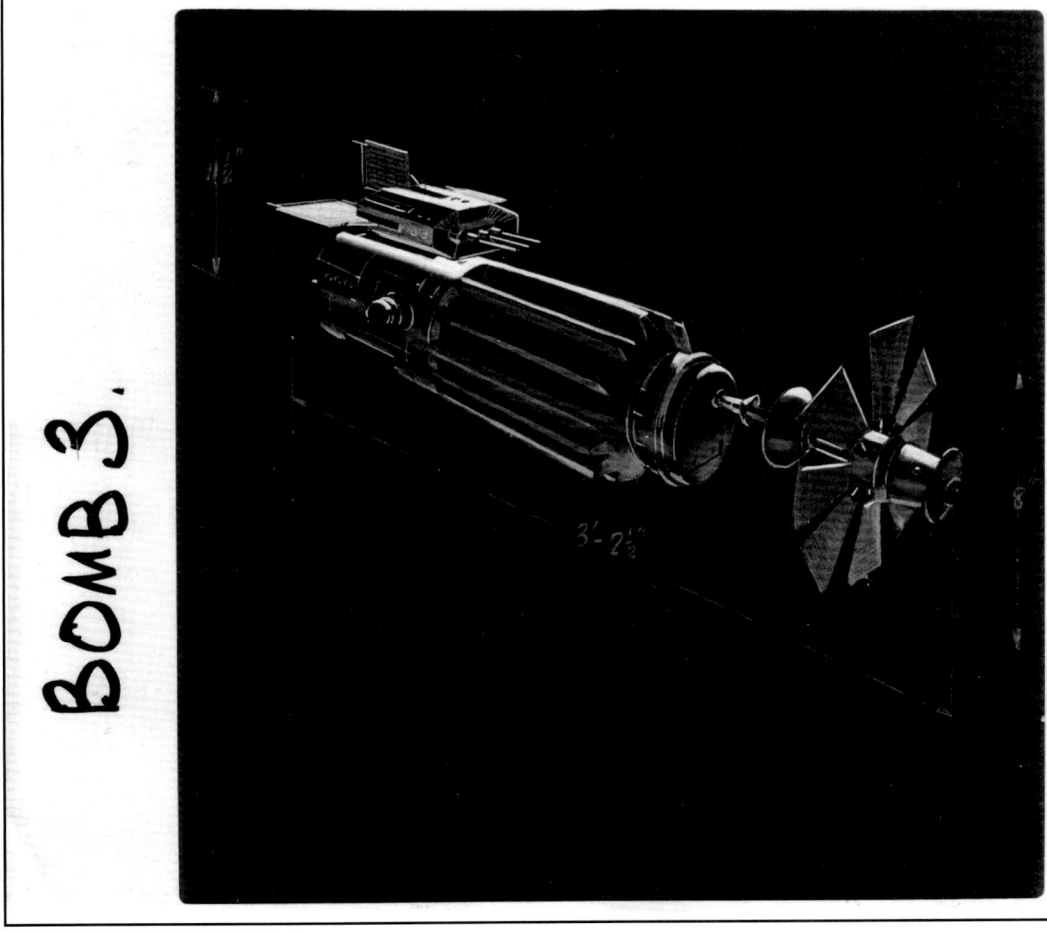

The story suggests that the U.S. Air Force has about a dozen of these satellite bombs in orbit. Here we see satellite #9 up close. It is quite small and was deployed by the NASA space shuttle *Orion* from its cargo bay. It is loaded with one nuclear space-to-ground or space-to-space missile. After launch, the satellite becomes permanently disabled. The satellite communicates with U.S. ground operations via its two antennae in the form of telemetry data.

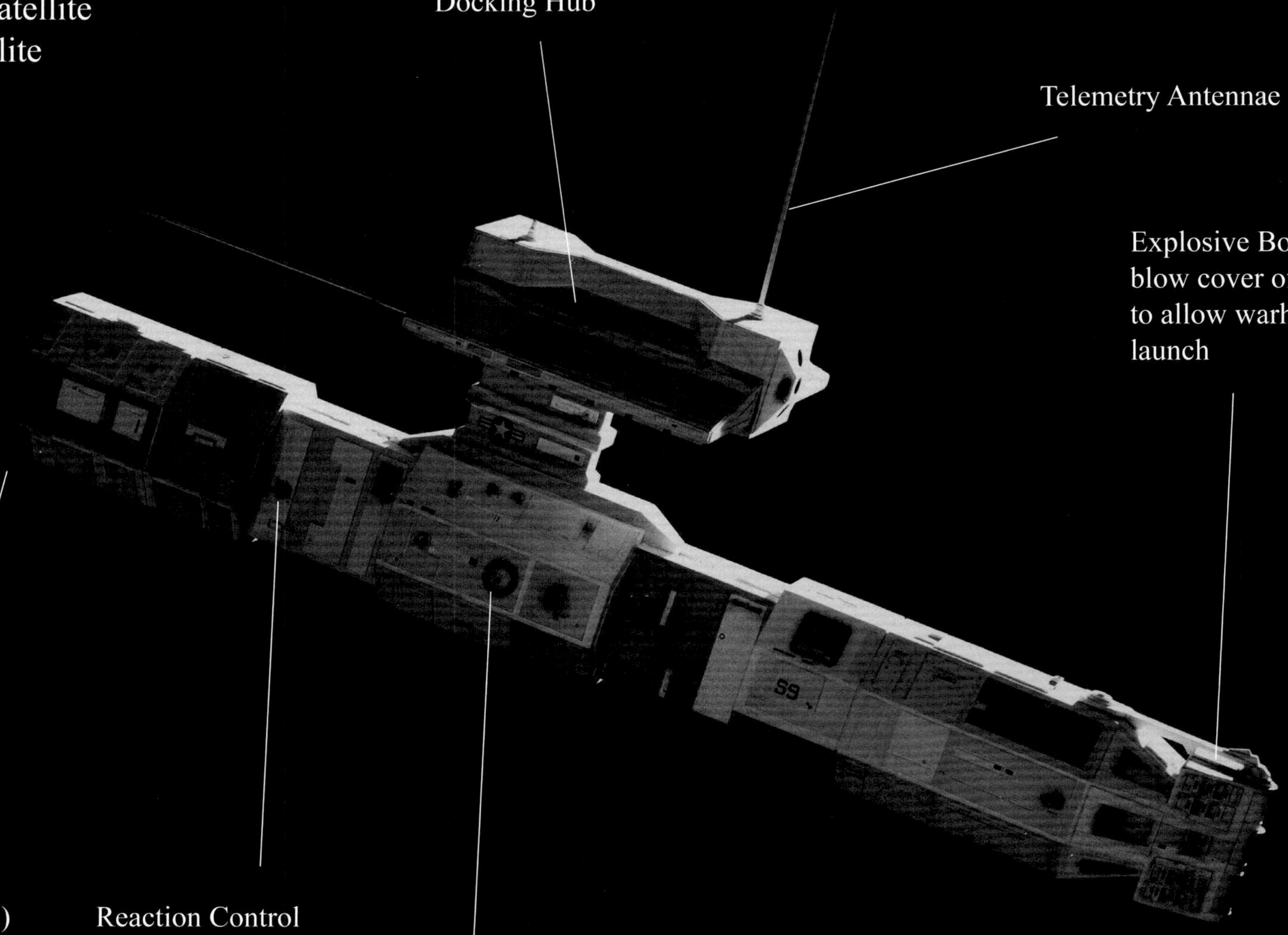

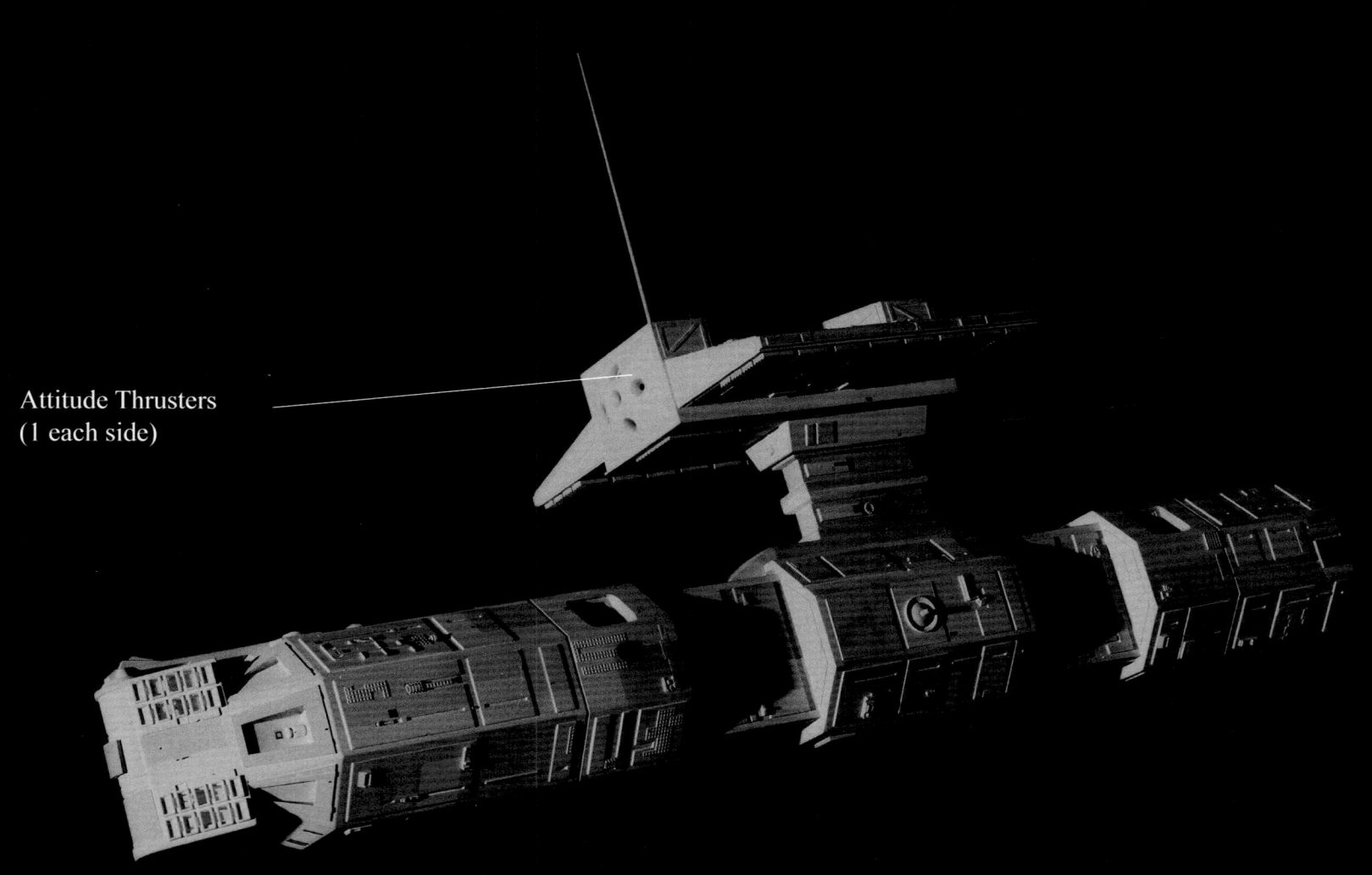

Docking Hub

Telemetry Antennae

Explosive Bolts blow cover off to allow warhead launch

Chemical Engines (5)

Reaction Control Thrusters (1 each side)

Docking Hub (1 each side)

Attitude Thrusters (1 each side)

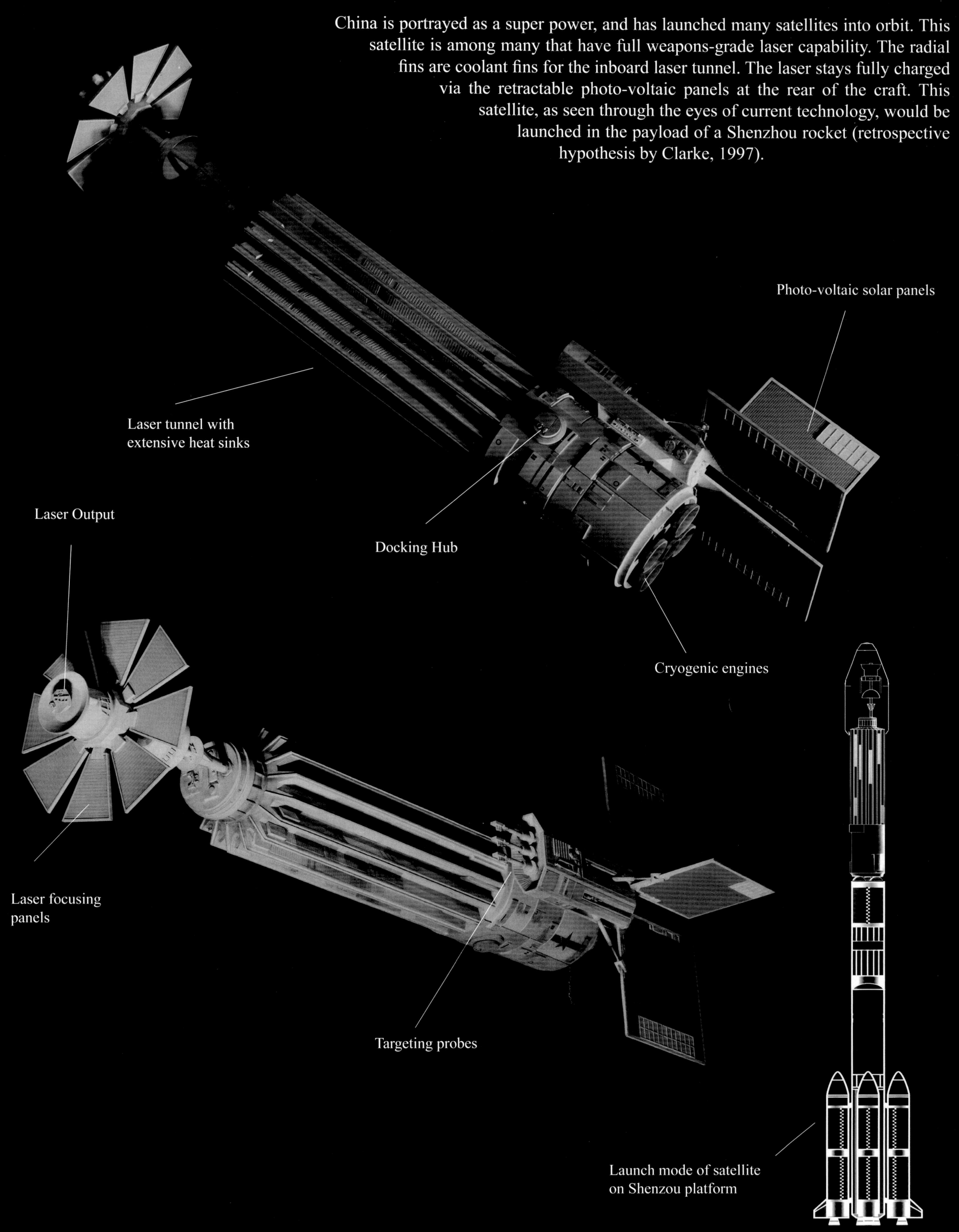

The French have always been at the cutting edge of aerospace technology. Here we have one of their satellites seen up close in orbit. It has many complex functions, including laser tracking, telemetry, and a large payload of five nuclear missiles. This satellite, as seen through the eyes of current technology, would be launched as the payload of an Ariane rocket (retrospective hypothesis by Clarke, 1997). The Ariane rockets are used by the unified nations of the European Space Agency or ESA. These rockets launch the bombs and satellites of countries including France, Germany, Britain, Spain, and Italy.

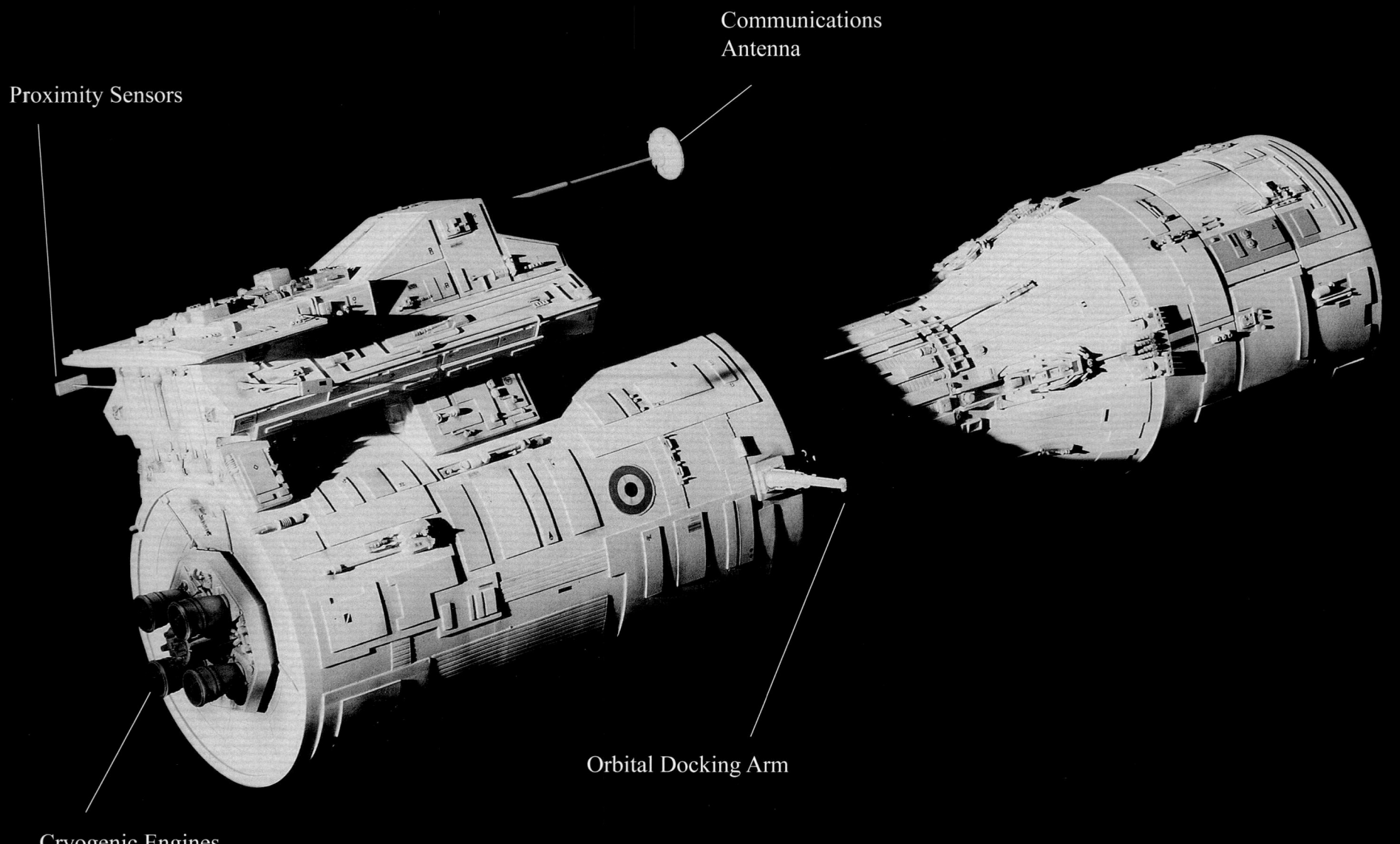

Proximity Sensors

Communications Antenna

Orbital Docking Arm

Cryogenic Engines

FLIR Targeting Window

Warhead Launch Point

Launch Mode of satellite on Arianne Platform

The German satellite seen here would also be launched from an Ariane rocket (retrospective hypothesis by Clarke, 1997). It is the largest satellite in orbit, and had to be assembled in space. The satellite is comprised of three pieces: the communications array, the cryogenic rocket engines, and the nuclear payload.

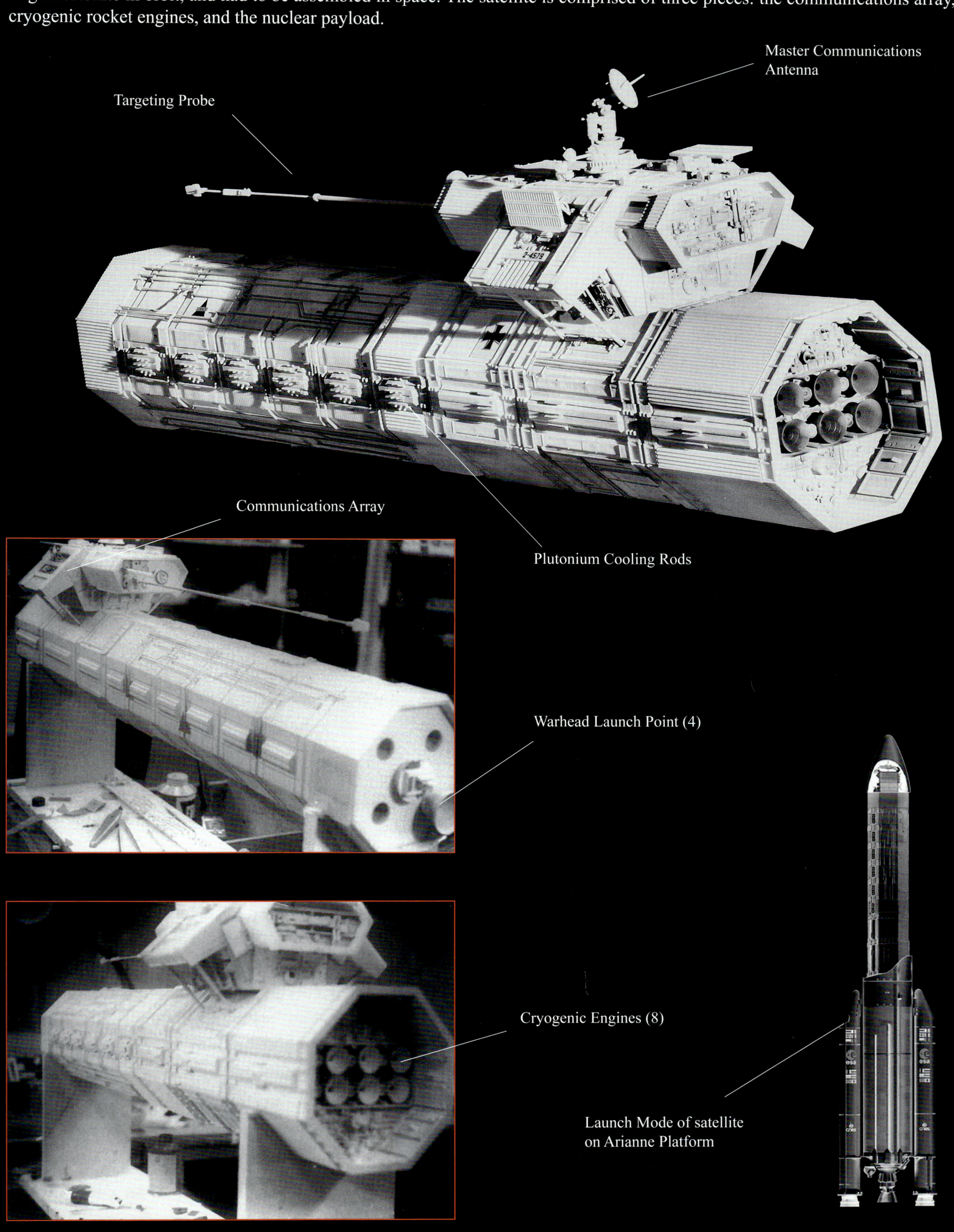

Master Communications Antenna

Targeting Probe

Communications Array

Plutonium Cooling Rods

Warhead Launch Point (4)

Cryogenic Engines (8)

Launch Mode of satellite on Arianne Platform

The Russian satellite seen here would be deployed from the cargo bay of the Buran-Energia shuttle (retrospective hypothesis by Clarke, 1997). It has ten nuclear warhead missiles in its body. The warheads are space-to-ground, or space-to-space missiles that can be launched from either end of the vehicle.

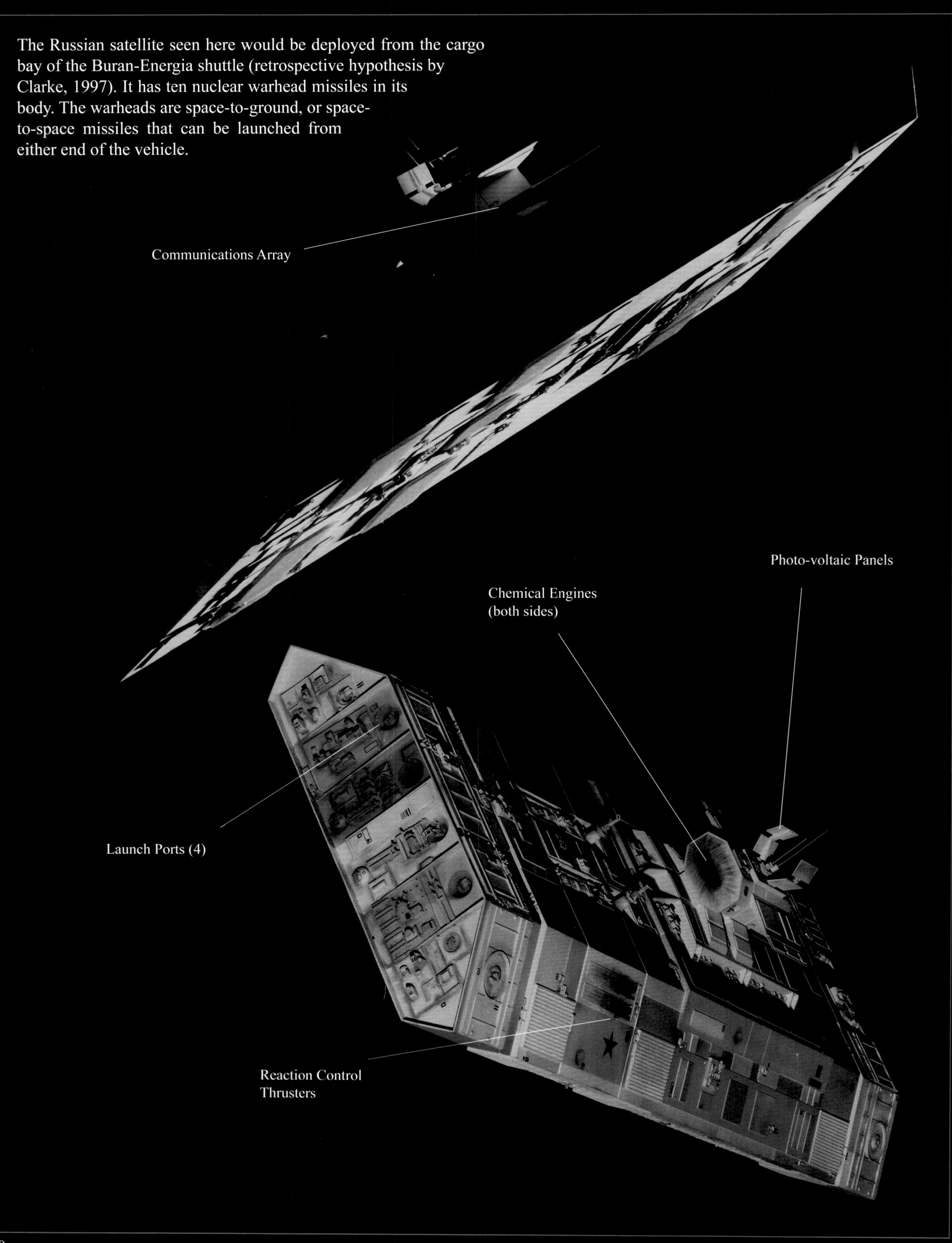

Communications Array

Photo-voltaic Panels

Chemical Engines (both sides)

Launch Ports (4)

Reaction Control Thrusters

The *Deep Space Monitor* satellite gathers information about asteroids, comets and meteors, and monitors the solar system for various objects orbiting the sun that may become a threat to Earth or its moon. This satellite has a large hi-resolution imaging telescope similar to the Hubble Orbiting Telescope. This is the satellite that discovered the second monolith (identical to the TMA-1 monolith found on the Moon) in orbit around Jupiter.

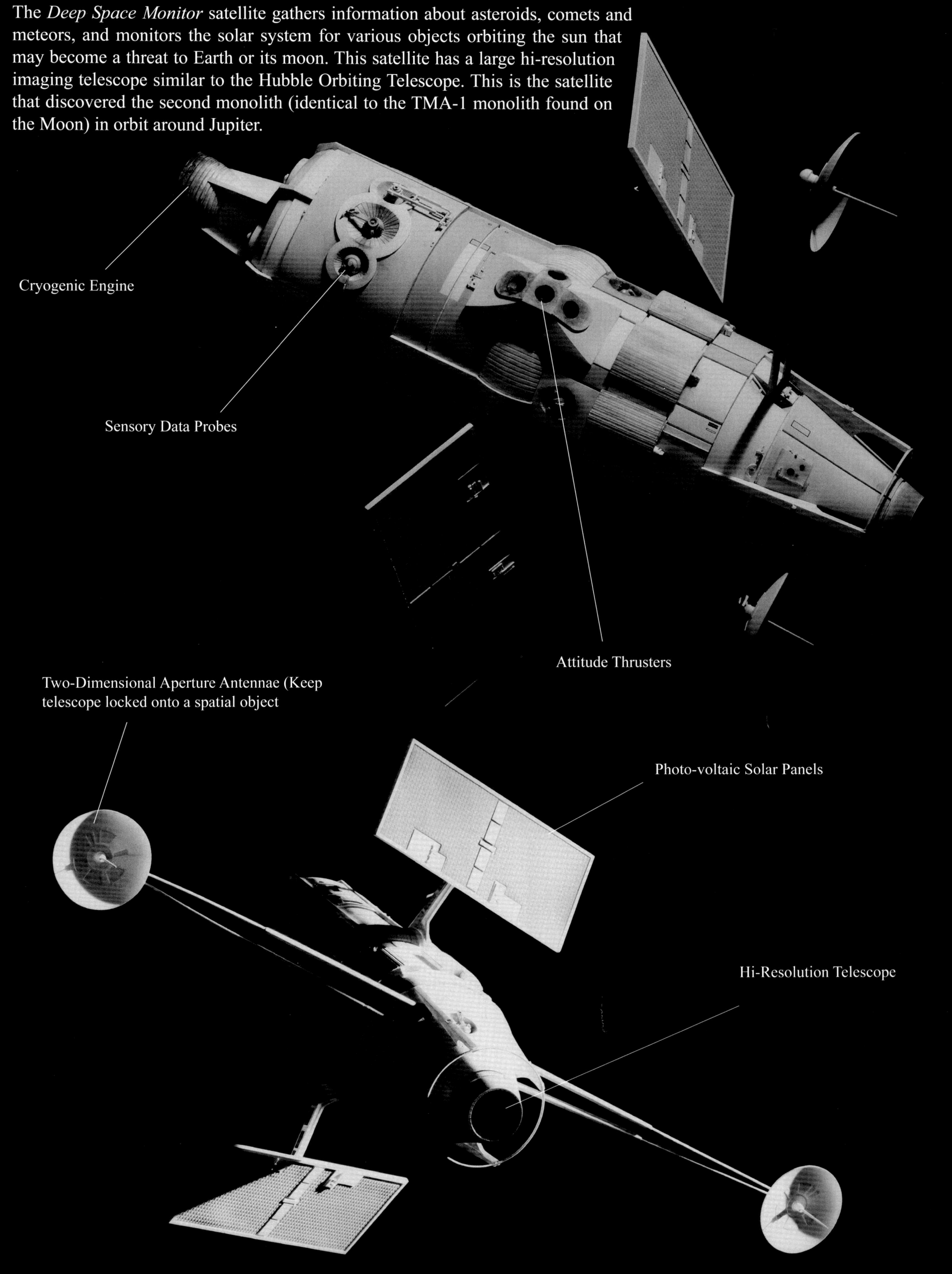

Cryogenic Engine

Sensory Data Probes

Attitude Thrusters

Two-Dimensional Aperture Antennae (Keep telescope locked onto a spatial object

Photo-voltaic Solar Panels

Hi-Resolution Telescope

The *Lunar Telemetry satellite* spends most of its time at a very low-level orbit and occasionally on the lunar surface mapping geophysical data, receiving and transmitting data of vehicles traveling around the moon. This satellite can be remotely guided to take off and land anywhere on the lunar surface or out into space. This is the satellite that discovered the Tycho Magnetic anomaly (TMA-1).

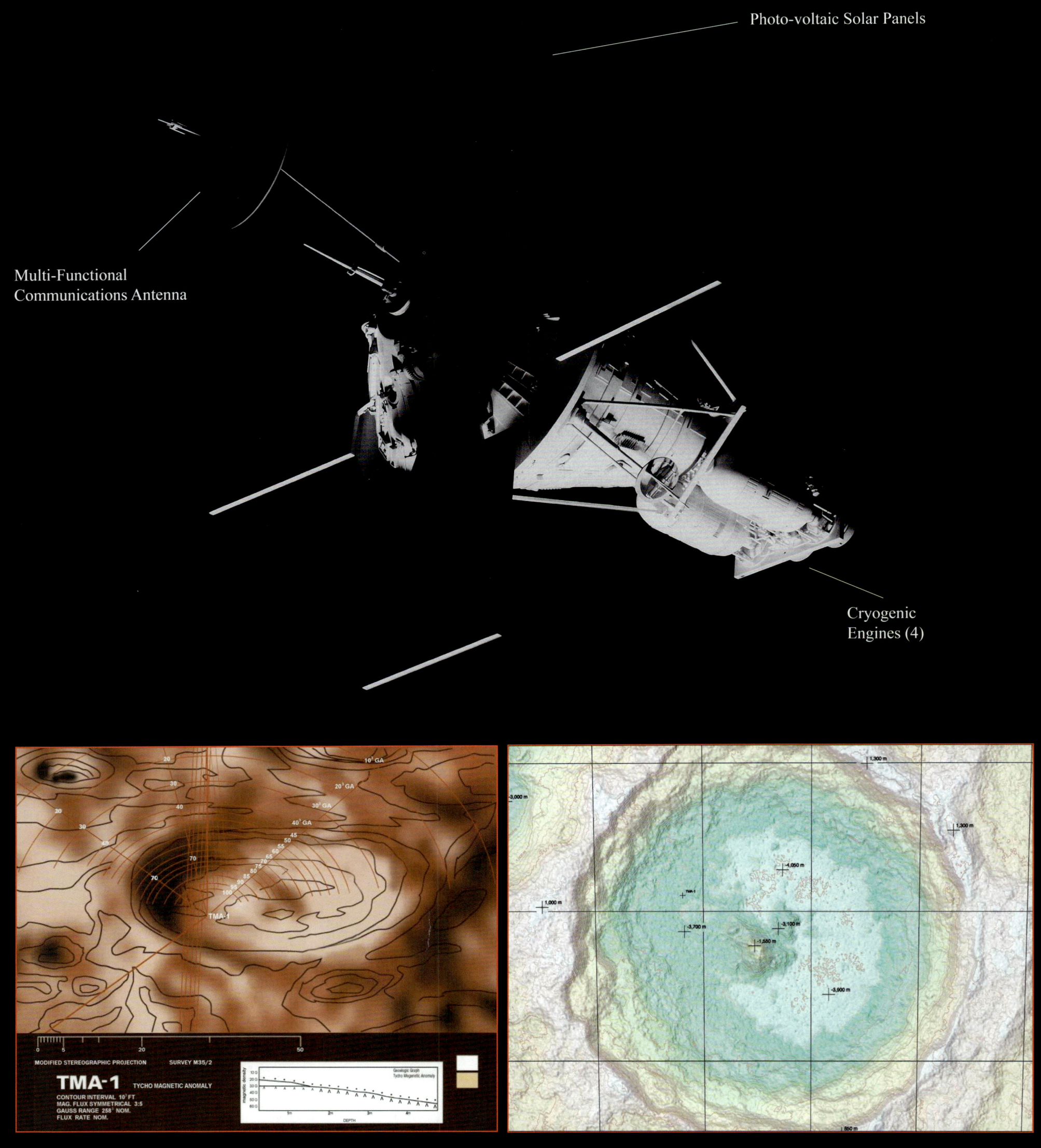

Examples of data received by satellite (Photo credit NASA)

The *Manned Observation satellite* (Cytherean Station One) which orbits Venus has three to four occupants inside its rudimentary enclosure. Its primary purpose is as a special space laboratory for unique experiments or observations of atmospheric and surface phenomena. Functions included dropping probes into the harsh Venusian atmosphere for geophysical analysis. Occupants stay no more than two weeks at a time.

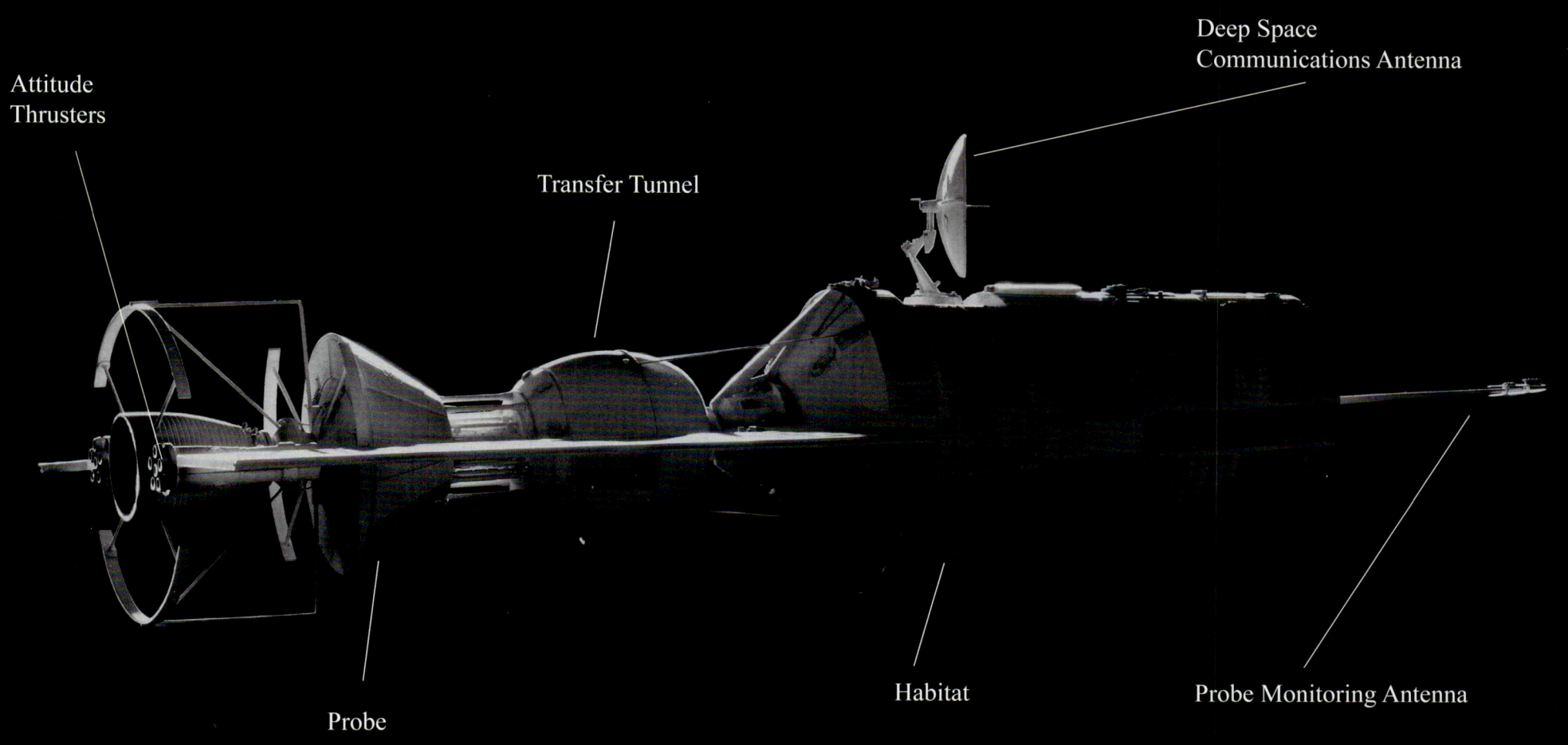

Cytherean Station One in low orbit over Venus. Note the strong resemblance of this to Skylab 1.

The Large Data Communications satellites (such as Intelsat VIII & X) are in high inclination earth orbits. They act as telemetry satellites for all types of commercial data transmitted to, or received from, Earth. This includes telephonic, TV, and GPS information, as predicted by Clarke in the mid-fifties.

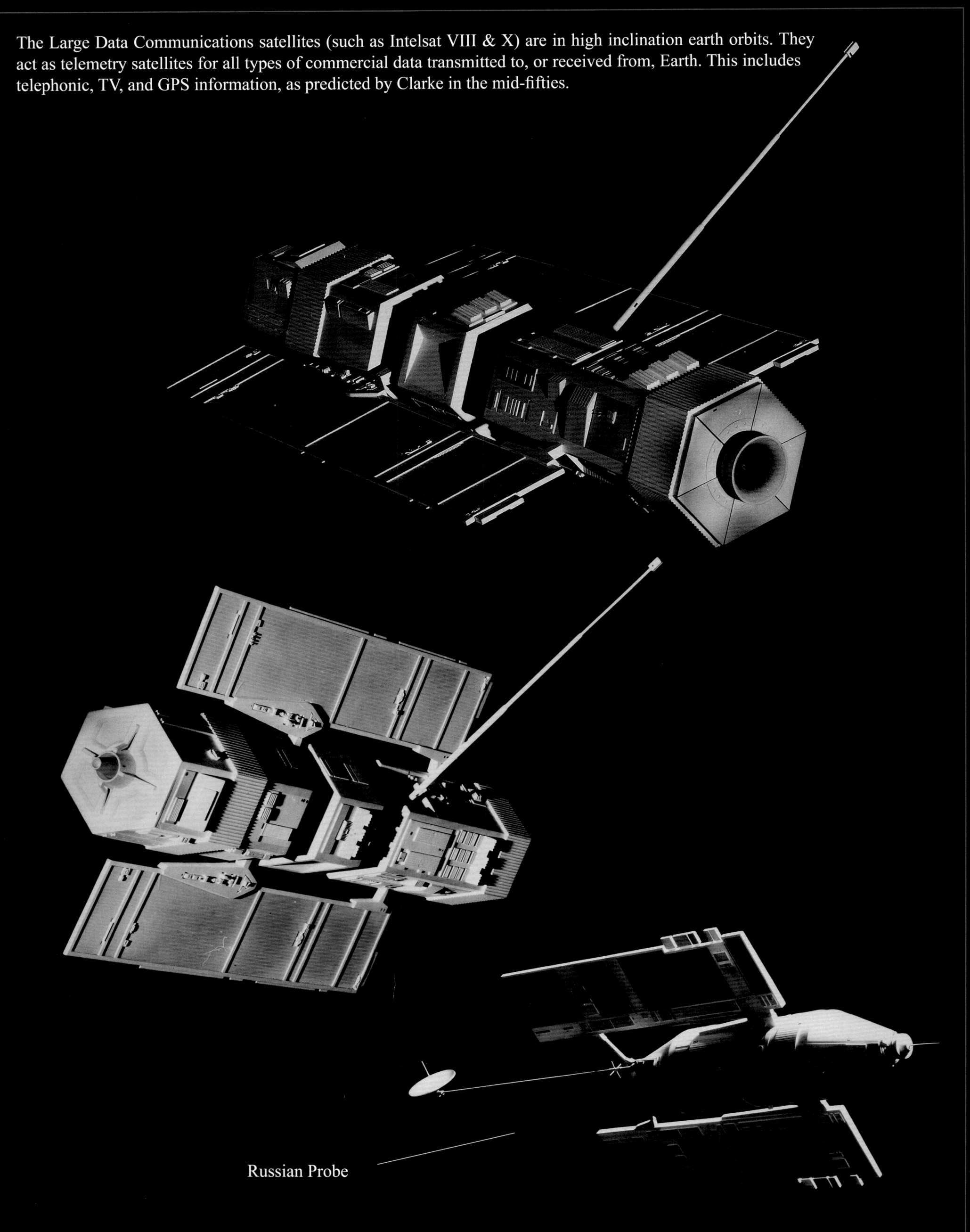

Russian Probe

THE ORION III SPACE PLANE

Orion III is a one and a half stage 175-foot long commercial shuttle, serving routes between Kennedy Space Center and the Space Station, consisting of expendable propellant tanks attached to a reusable, integral hull of high mass fraction. Boosted takeoff occurred from an inclined ramp and, after aerodynamic entry, it would land on a conventional horizontal runway. The vehicle can carry up to 30 passengers, a pilot and a copilot/engineer/navigator. On-board systems incorporate a variety of sophisticated electronic computer, guidance, docking, communications, and other equipment, much of it developed with the assistance of IBM, Honeywell, RCA, General Electric, and other firms. IBM's computer panels, for example, handle primary functions such as navigation and docking. Data is processed routinely and automatically, but occasionally there is a call for "priority interrupt." The program is loaded and the appropriate button, e.g., "propulsion", is depressed. On the screen the words "propulsion program loaded--indicate requirements" might appear, followed by a series of major category statements, e.g., propellant analysis, predicted propellant utilization, and tank pressures. The astronaut would then depress the appropriate button on the keyboard, e.g. the number 2, and would enter "require predicted utilization for next 3.8 hours based on (1) current rate, (2) revised rate if No. 3 engine thrust is increased 0.96 per cent." This would be a "routine" under the "predicted utilization" subcategory, illustrating man-machine interface as the computer assists the crew throughout the sequence.

IBM's main on-board computer is supported by backup computers. A light goes on when the main computer is either overloaded or cannot process the desired requests. It then automatically addresses backup computer No. 1, whose indicator light goes green if it can accept the request. If not, backup computer No. 2 is interrogated.

Other major elements of *Orion III* computers are tasks in execution, systems mode, integrated checkout, narrative instructions, digital entry, program status (display energize, memory load, error check, etc.), various data review modes, instrument self-test, condition display, diagnostic, program mode (e.g., conversation mode "keyboard", conversation mode "oral"), message assembly, readiness checkout sequence, service equipment control verification (with selected sequences), alarm conditions (with selected stations indicated), and verification routine. For docking with the Space Station, the flight computer is linked to a FLIR (forward looking infra-red) tracking system in the nose of the vehicle. The computer in both the Space Station and *Orion* then link via telemetry. This allows the *Orion* to lock onto the rotational speed and relative position in space precisely.

The Orion take-off stage was inspired by the Feiseler aircraft company's early V-1 pulse-jet designs of WWII which took off on an inclined ramp. The 'piggy back' design of Harry Lange seen here, was derived from Willy Ley's* concepts of earth-to-orbit space travel in the 1950's. These concepts engaged the general public, science fiction film designers, and eventually NASA.

The complex computer systems in the cockpit. Seen here is the Space Station being digitally tracked.

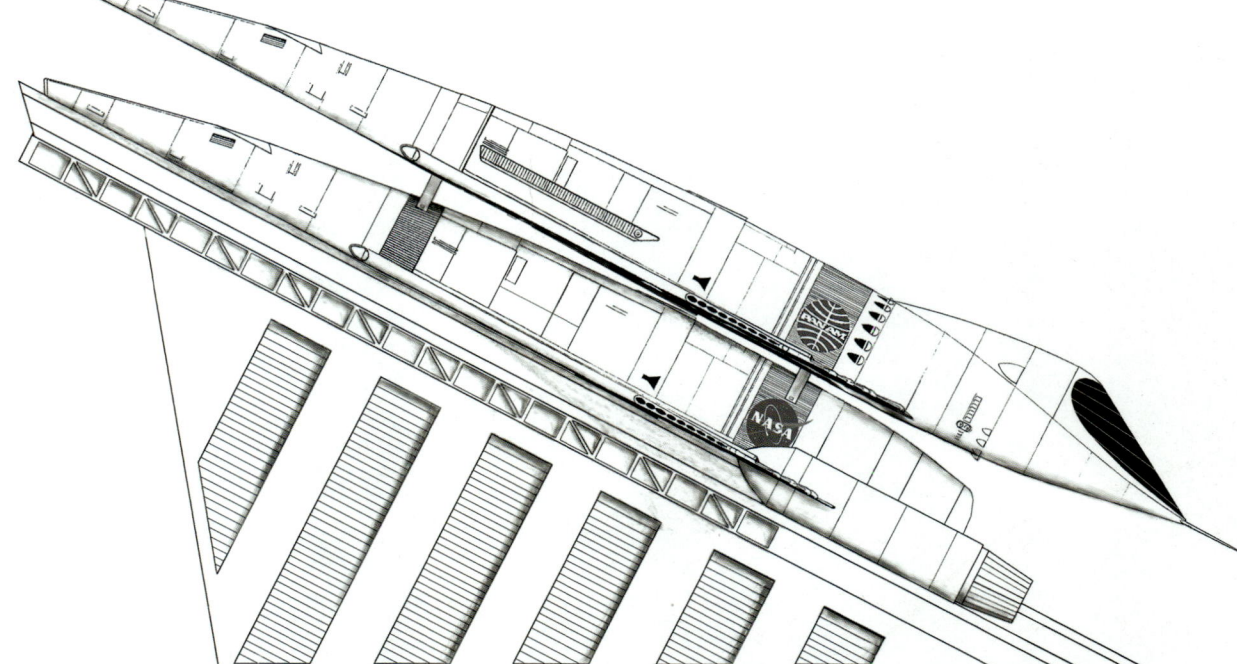

Orion III atop Orion I booster. Take-off mode. Carries Orion III, achieves Mach 14 then disconnects and sails back to Earth.

* Willy Ley: Space sciences academic and science fiction popularist of the 1950's.

15

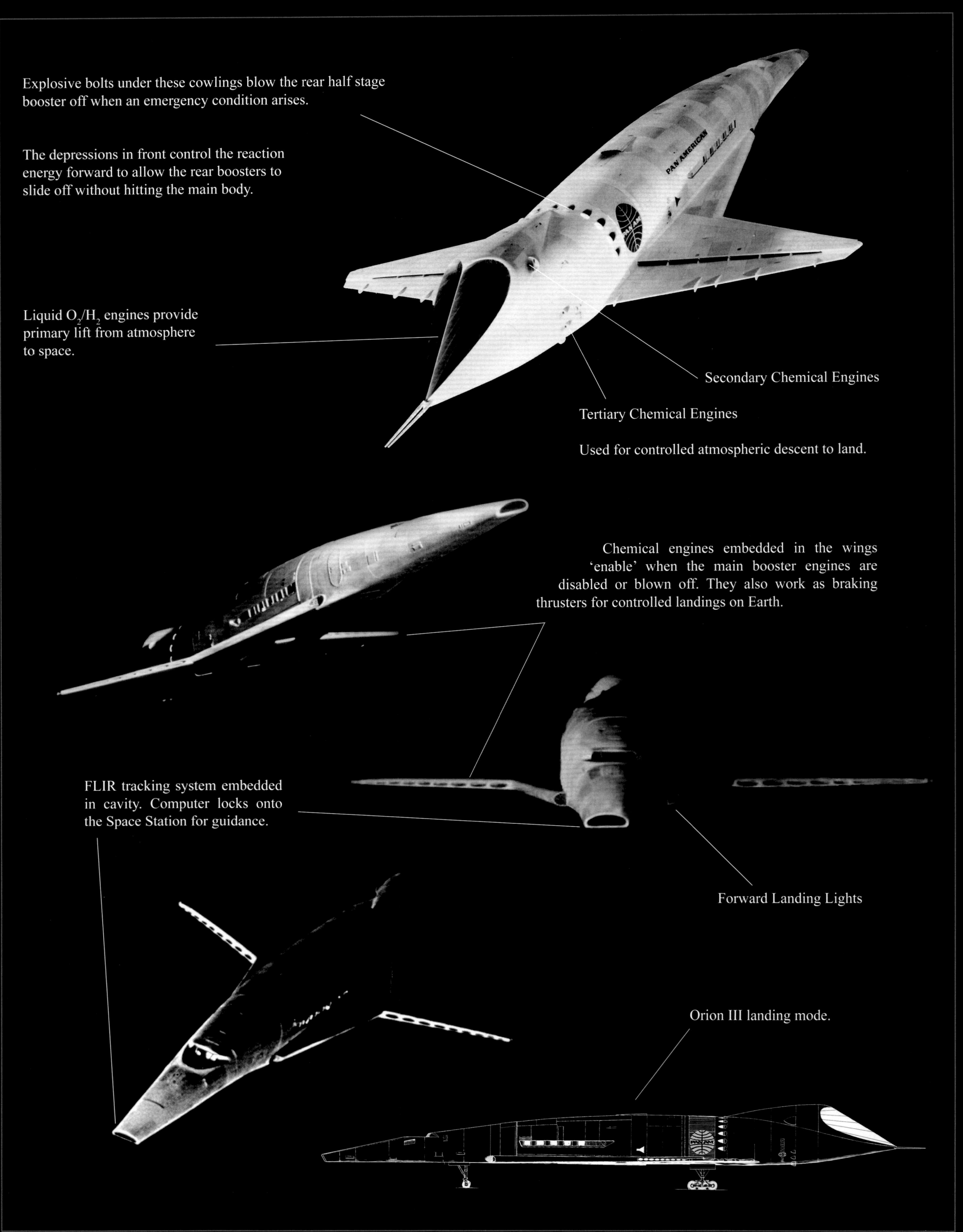

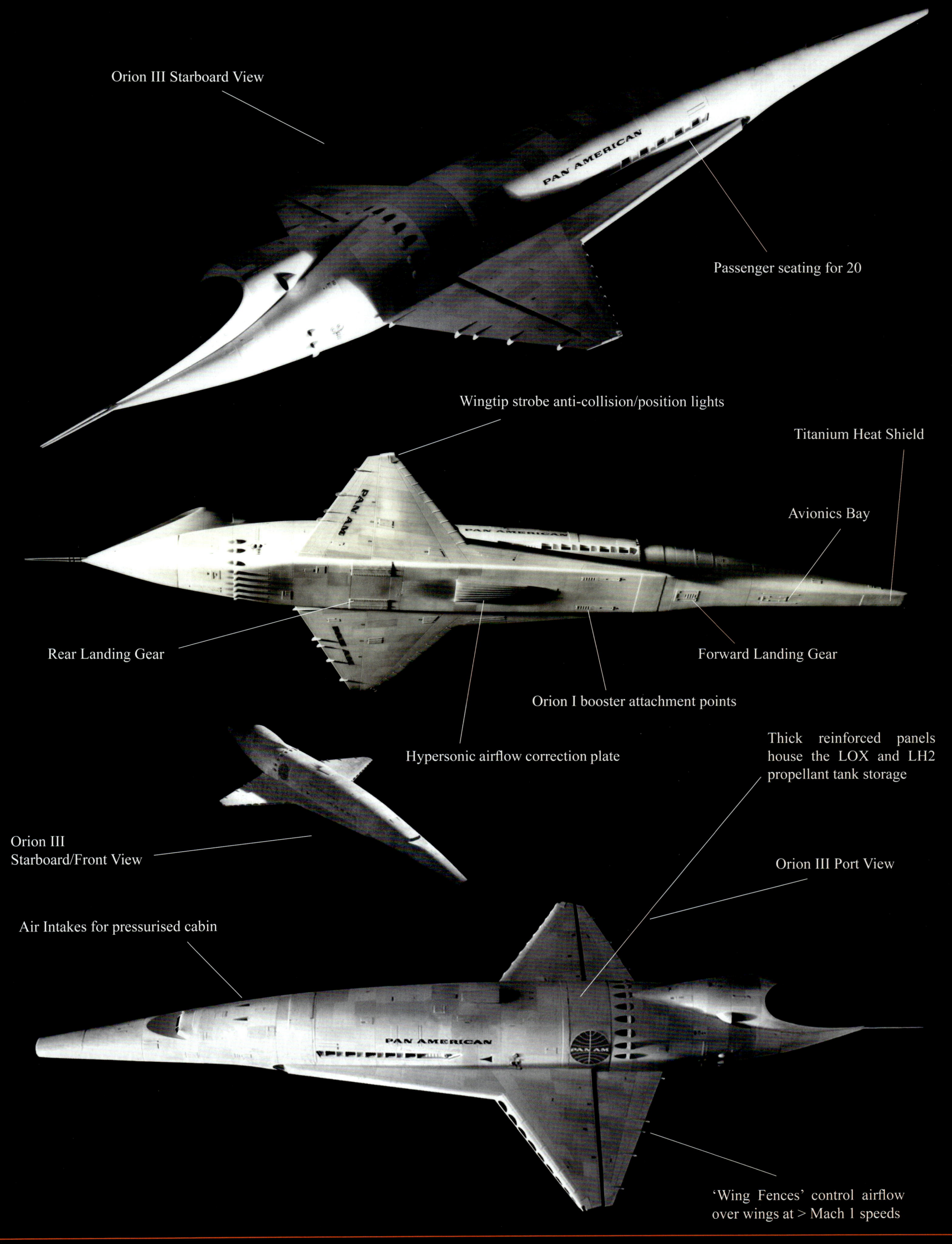

Orion III schematic – starboard and rear views Final design (seen on previous two pages), demonstrates the 'engine hump' to be slightly wider, and surface paneling changed based on construction. A re-entry heat shield slides over the passenger windows (not visible in drawing).

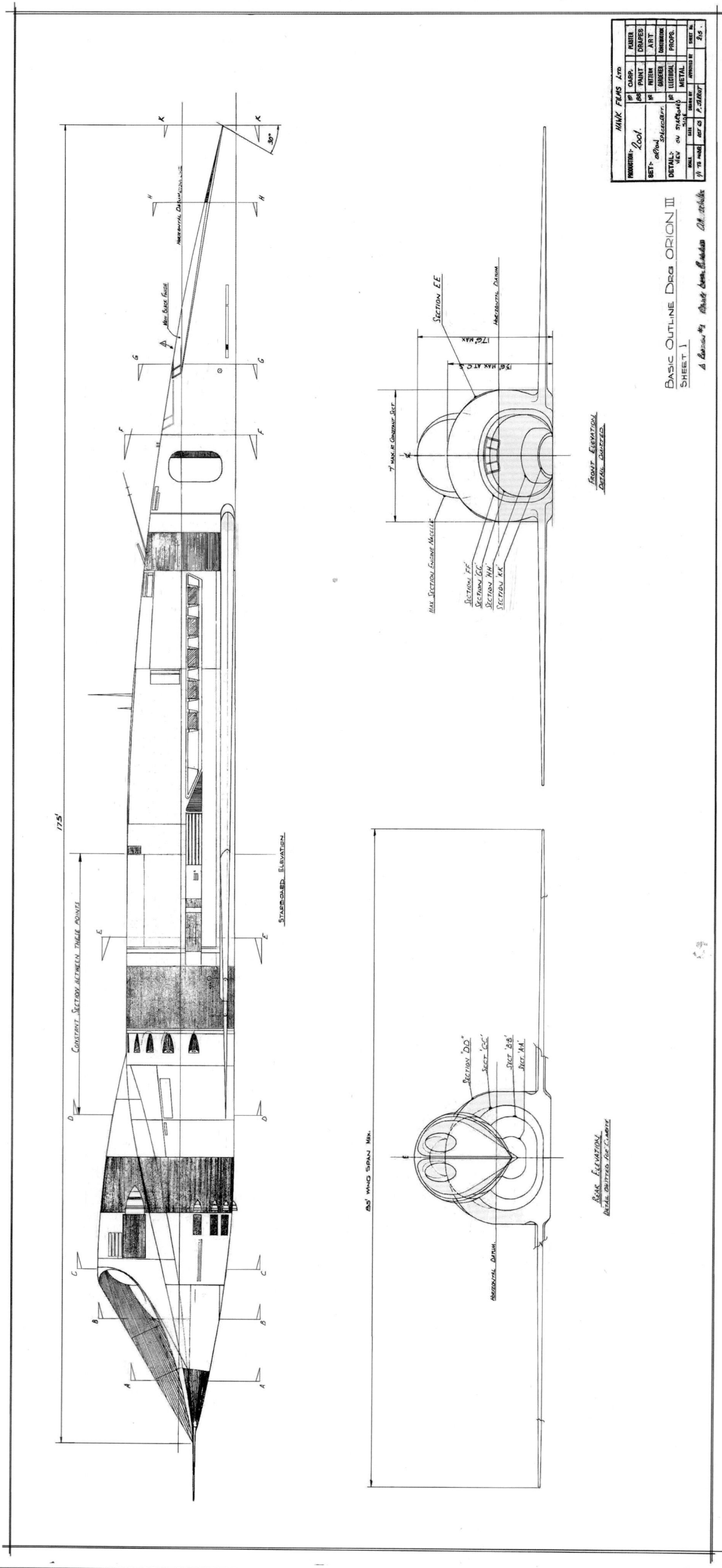

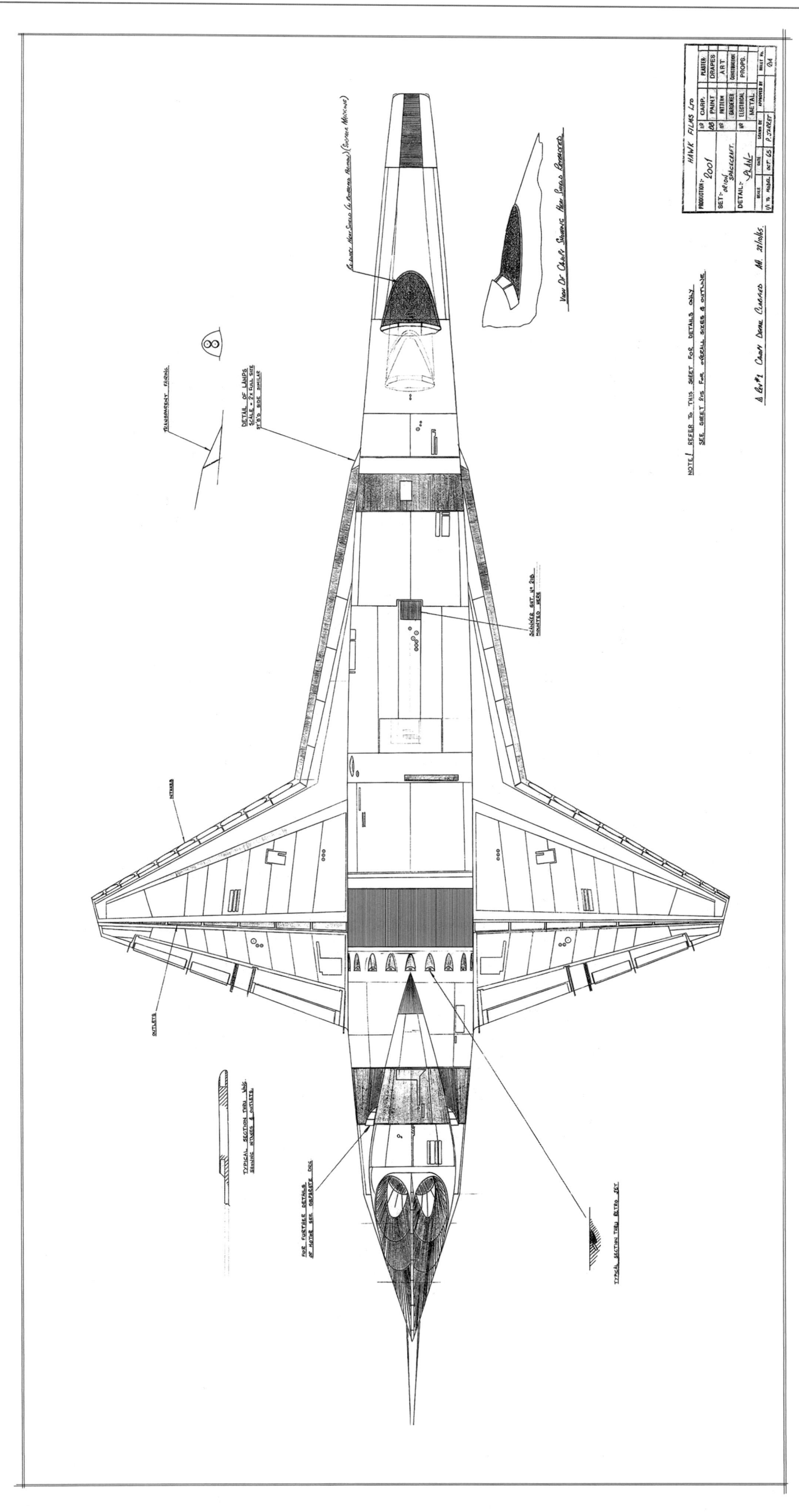

Orion III schematic – Plan view. Horizontal surface directly in front of cockpit window can be raised and retracted when leaving or re-entering Earth's atmosphere. Design characteristics sourced from the North American XB-70 Valkyrie front cockpit window.

Fred Ordway and Harry Lange consult with each other on final details of the passenger vehicle.

Masters, Lange, Ordway discussing the Orion designs.

Fred Ordway and three consultants from Hawker Siddeley Dynamics discuss technical details of the cockpit and passenger area.

The Parker Atomic Pen has a tiny nuclear isotope in it to provide heat. The ink, solid at room temperature, is melted by the isotope and supplied to the nib. Ink flows by engaging this button.

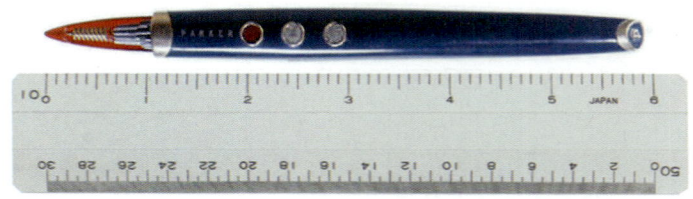

A stewardess grabs a passenger's pen that inadvertently floated out of his pocket in the zero-G conditions of space.

Flat panel TV's are now the norm on the backs of passenger seating in airliners.

TITOV V SPACEPLANE/SHUTTLE

Most likely built by Buran, this is Russia's version of the *Orion III,* and is 200 feet long. It has virtually identical functions other than it can carry passengers *and* payload to space and return to earth like a conventional passenger jet. The Russians have historically proven that reliability is a function of continued regular use of aerospace vehicles, thus contemporary conjecture is that the Energia Booster would launch this spaceplane into orbit (very similar to the US Space shuttle/booster).

Titov V on approach to the space station. Wing engines similar to the Orion are used for controlled descent to land on Earth as a conventional jet.

Alternate view of the Titov V. Note: rear engines are missing in this photo. This is an early mock-up that was incomplete.

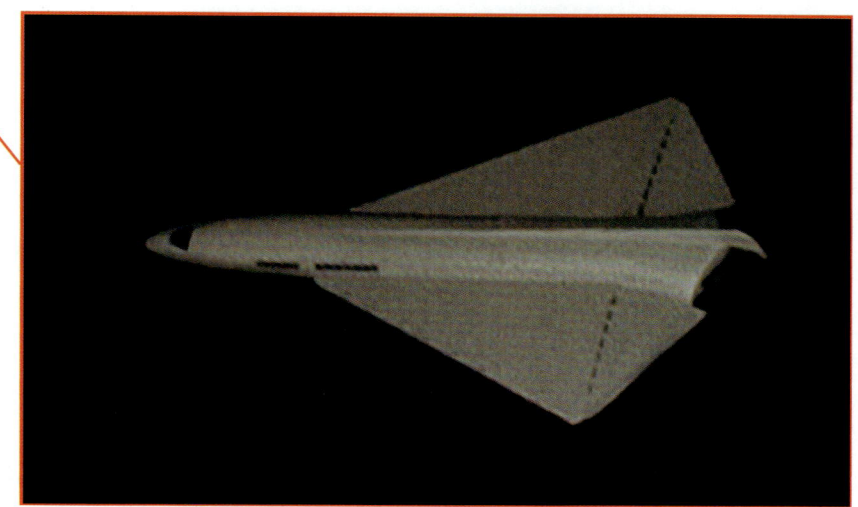

Russian crew takes time to relax in the Space Station before departing back to Earth.

21

SPACE STATION V

The huge 1000-foot diameter, spinning station in space, is of modular design, and is inclined at approximately 50 degrees to the equator at an altitude of nearly 300 miles. Docking takes place in the rectangular portal in the middle, and people are transferred to the rim by elevator. Gravity forces at the rim are 0.2 *g*. All docking and cargo handling takes place in the core in a micro-gravity environment. Crew, technician and resident workers live in pleasant rooms for rotated periods of 90 or 180 days. Assembly workers would come and go in accordance with requirements. Horizon scanners, star trackers, and rate gyros were installed on the station to provide necessary sensing data. Attitude stabilization is assured by control movement gyros and reaction control thruster systems.

The scenes in the film showing the Orion III next to the Space Station have been scaled down to make the Space Station appear much larger than it actually is.

The 150-foot space station set built for the film was sized at a circumference of 3000 feet. It was decided to lower the floor angle so actors would not slide down the floor due to the steep angle.

The modular design has two *Orb Discs*:

ORB DISC 1
Outer Rim (B)

Reception and Customs
Banking and Credit Offices
Hotel Reception and Shopping Plaza
Bar, Coffee Shop & Lounge
Vision Phones Computation Laboratory
Restaurant
Kitchen and Storage
Station Personnel Dining
Orbital Meteorological Observatory
Shop and Repair
Astronomy and Geophysics Laboratory
Photo Laboratory
Instrument Laboratory
Passenger Telescopes
Space Science and Technology Supervisory
Scientists Offices
Command and Control Deck
International Operations Office
US Astronautics Agency Offices

Inner Rim (A) Hotel Block A

Hotel Block B
Hotel Block C
Cinema and TV
Conference and Display
Communications
Gymnasium
Hospital, Medical Laboratory and Human Factors
Research
Out Patients
Administrative Offices
Station Quarters A
Station Quarters B
Storage and Supplies Crew Access Only (Red Badges)
Station Quarters C
Engine Room (Electrical power distribution, Atmospheric gas supply, Environmental Control)

ORB DISC 2
Outer Rim (B) Astrobiology Laboratories

Astrochemistry Laboratories
Space Physics Laboratories
Restricted area -Green Badge Only *Inner Rim (B)*
Astrobiology Laboratories
Vacuum Research Laboratories
Space Physics Laboratories
Restricted area -Green Badge Only

Two views of Space Station V in orbit. The eight arms support the habitat rims and also have circular elevators that transport personnel and equipment to the rims.

Ordway, Ernest Bevilacqua and Dr. Richard Leakey* in the elevator during final construction. *Bevilacqua was the consultant from IBM. Leakey was the anthropology consultant for the 'Dawn of Man' sequence at the beginning of the film.

Kubrick and Ordway take a break!

Kubrick and Ordway in the main corridor of the inner rim.

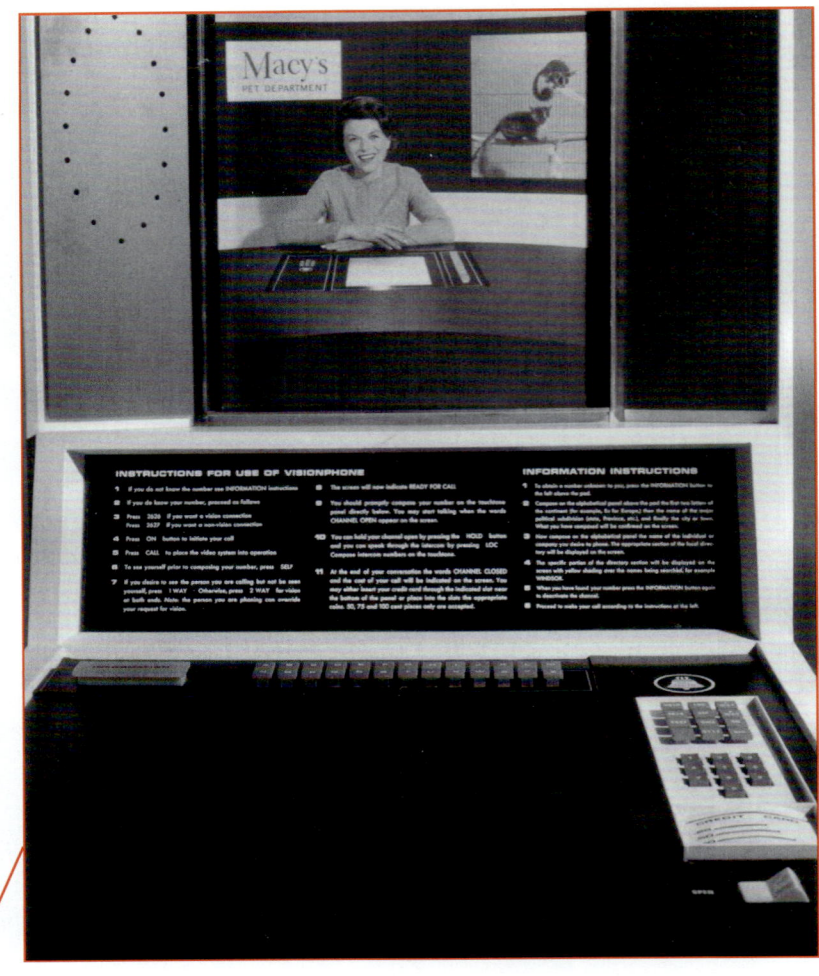

Picture Phones: To obtain a number, one first presses the information request button and then composes on the alphabetical panel the first two letters of the continent – for example, EU for Europe, next the name of the major political subdivision, and finally the city or town. Everything already composed is confirmed on the screen in front of the customer. By using the alphabetical panel, the name of the individual or organization is requested. Almost instantaneously, the appropriate section of the local directory is displayed, the name being hunted indicated by a yellow shading. When making the call itself, appropriate buttons can be depressed in accordance if one prefers a vision or a nonvision connection. A 'call' button places the video system into operation, and a 'self' button permits one to see oneself prior to composing the number. There is also provision for both one-way and two-way vision. However, the person being called can override a request for two-way vision. When the screen indicates 'ready to call,' the number is composed on the touchtone panel. When the words 'channel open' appear on the screen, conversation can start. At the end of the conversation, a 'channel closed' announcement appears, together with the cost of the call – normally chargeable by credit card.

A travel agent is available to service guests. Most travel agent offices had fiber-glass model mock-ups of the various airlines they represented up until the year 2000. In the foreground, various moon rocks are available for sale.

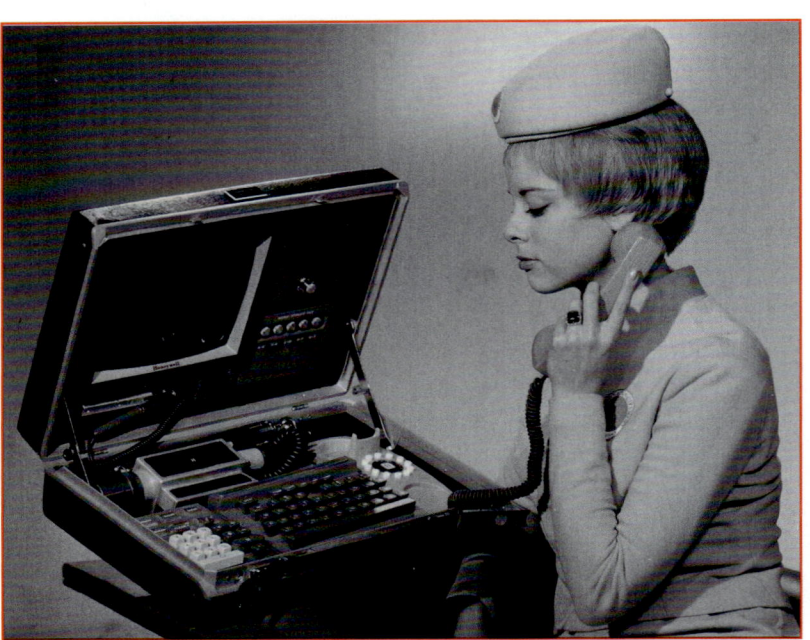

An employee of the Space Station demonstrates how to use a state-of-the-art Laptop computer designed by Honeywell. The unit features a GPS linked telephone and built-in printer. Such devices are available to guests on the Space Station.

A Coffee Bar is situated in many hallways of the Space Station. By depressing a series of buttons, one can choose for example: Coffee-Black, Coffee with cream, Mocha, Cappuccino, etc. The cup of fluid is removed from the dispenser by her right hand.

ARIES 1B SHUTTLE

Aries 1B is a 42-foot diameter space-to-moon, moon-to-space shuttle, designed to transport up to 24 passengers and four crew members between the Space Station and the moon. It is identical to the *Aries 1A* (without passenger windows) whose sole function is to transport cargo to supply the *Clavius Moonbase*. The on-board computer systems and cockpit design are identical to the *Orion III*. A well-stocked galley supplied by Whirlpool houses a sophisticated automatic kitchen on the lowest deck by the entry door, to supply passengers and crew with snacks, meals, and drinks on the long journeys to and from the moon. A unique feature of the passenger area is two bathrooms. Again, supplied by RCA/Whirlpool. *Note: Whirlpool was chosen after the successful implementation of the feeding and waste systems on board the Gemini spacecraft in the 1960s.*

Aries lands at Clavius Moonbase with the addition of Reaction Control thrusters on four locations outside the main body, and returns to the Space Station via FLIR/computer tracking and telemetry identical to *Orion*.

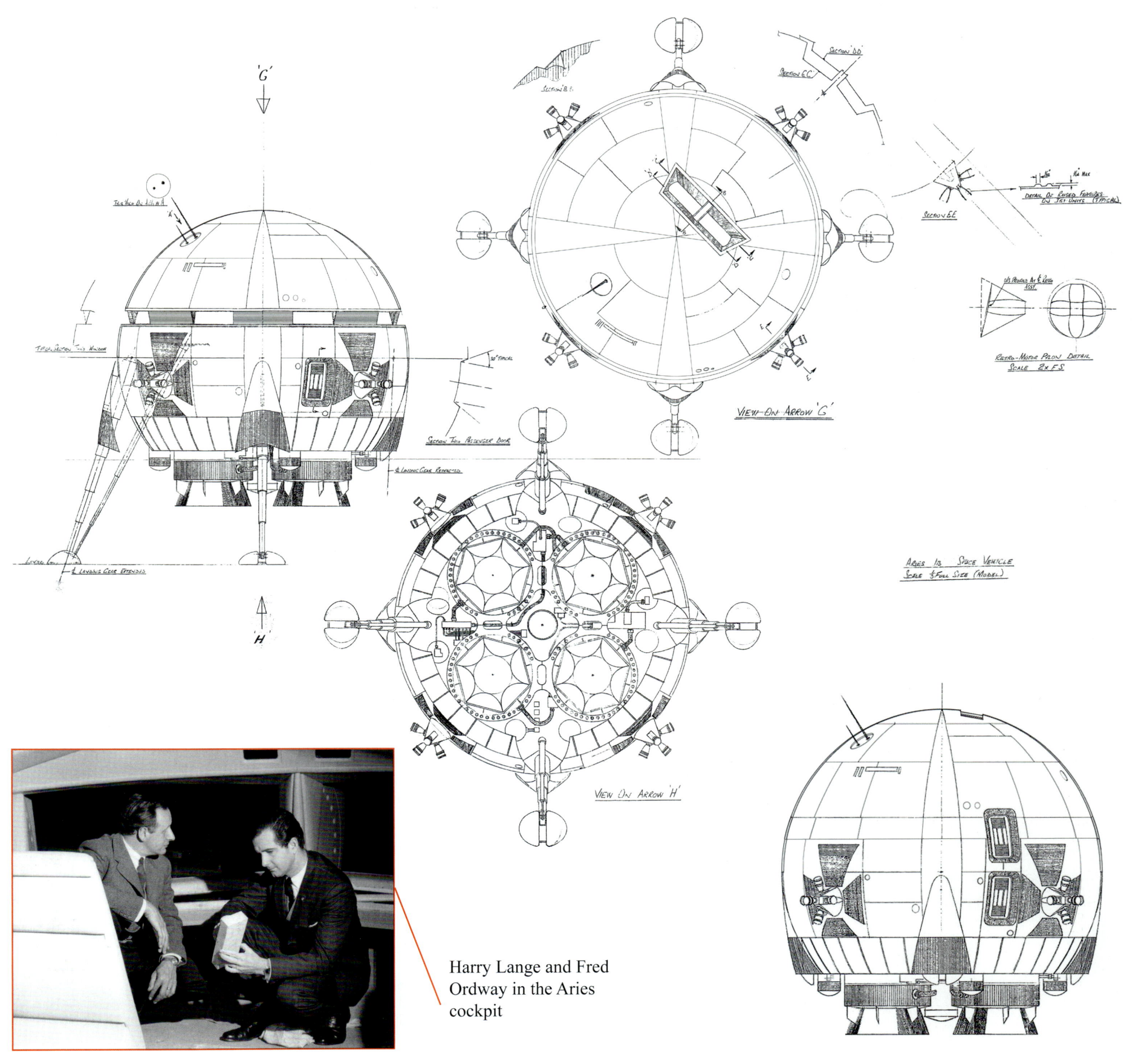

Harry Lange and Fred Ordway in the Aries cockpit

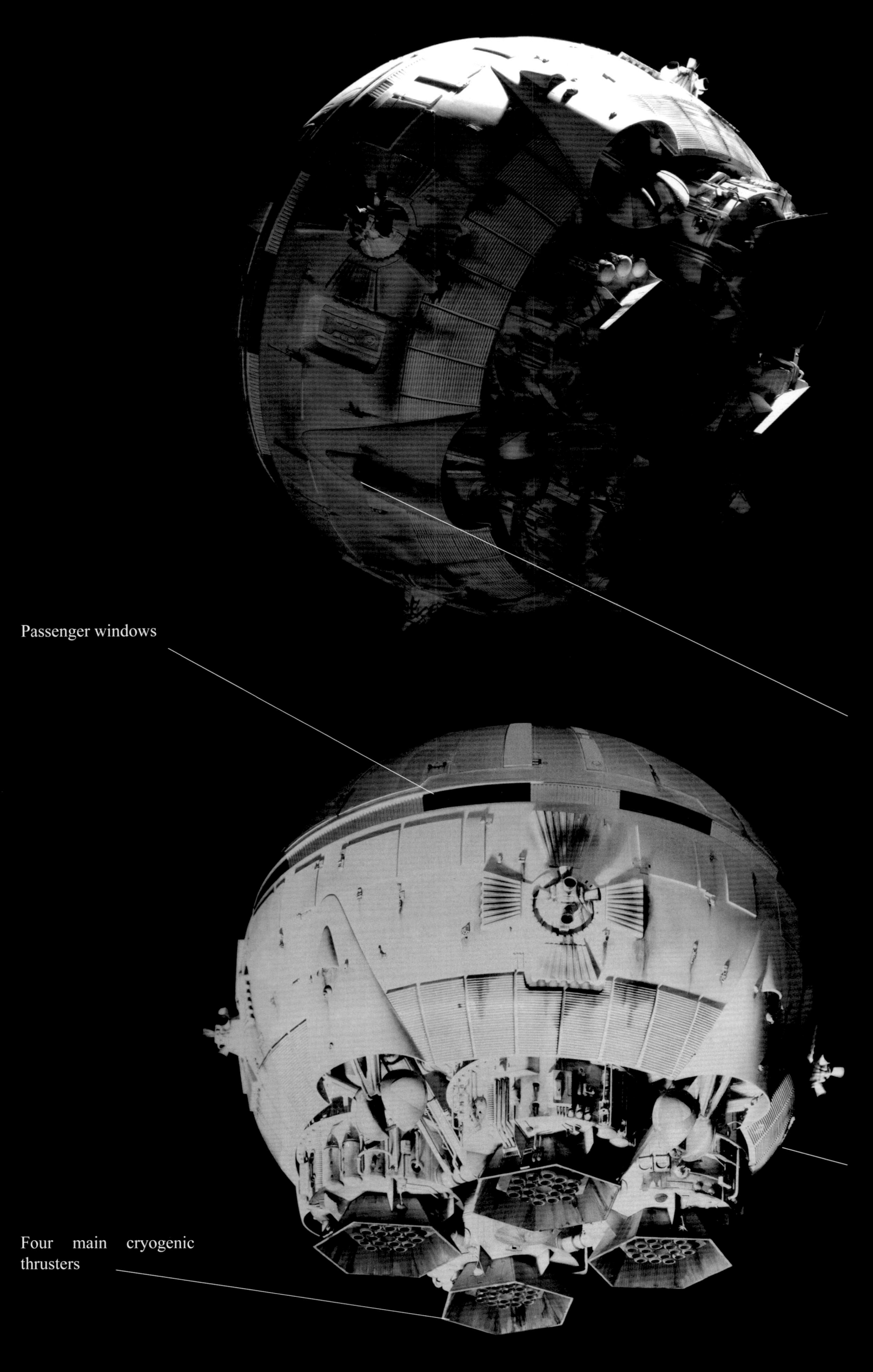

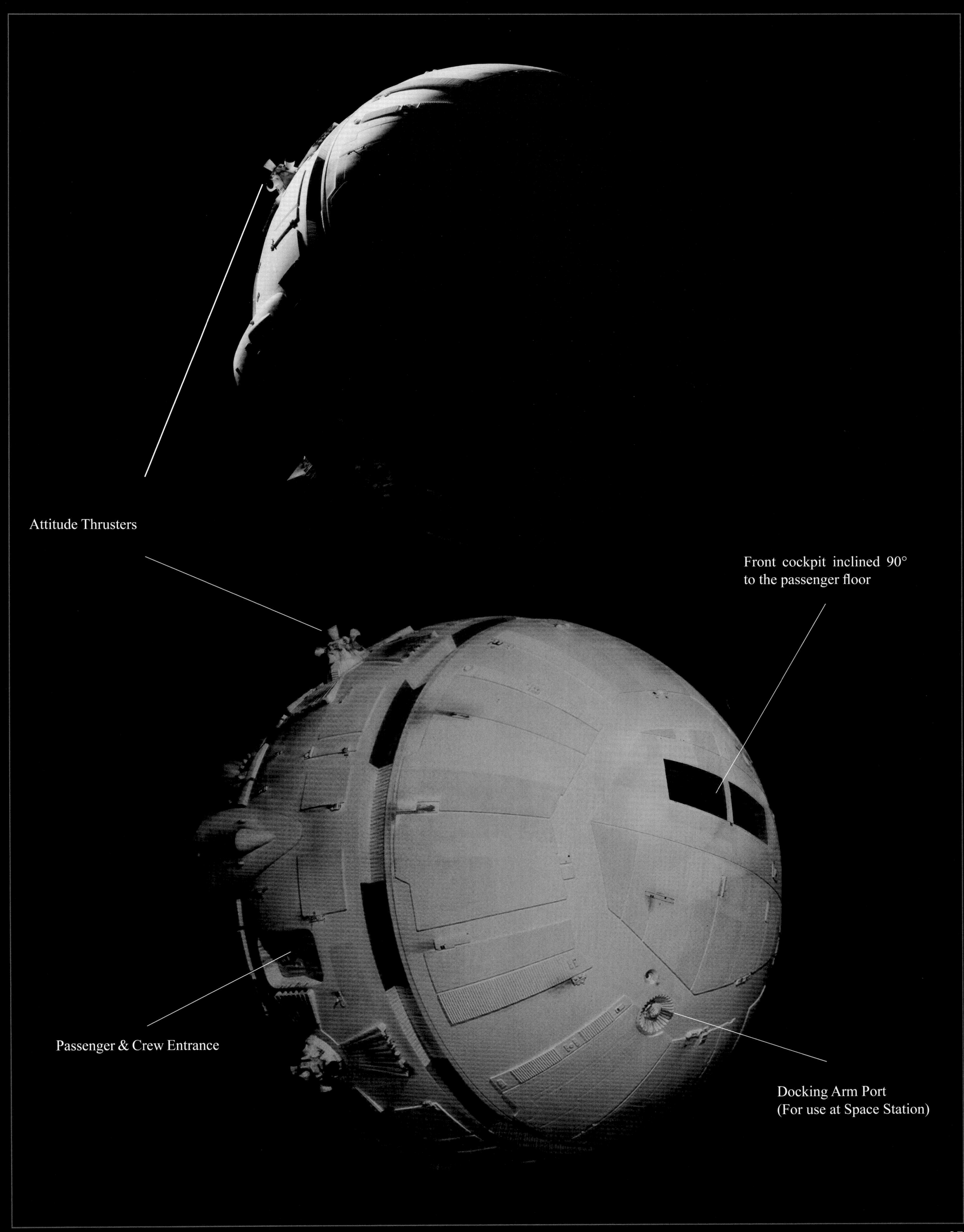

Attitude Thrusters

Front cockpit inclined 90° to the passenger floor

Passenger & Crew Entrance

Docking Arm Port (For use at Space Station)

The engines are of the cryogenic type LOX/LH2. Multiple nozzles on the four main engines allow for stable movements and blast conditions. The large cylinder directly in the center of the four main engines contains all the avionics and gyroscopic devices for controlled landings and take-offs from the moon. The many conduits around the engines carry ESPS signals, sensor data and optical data to the cockpit.

The sides of the sphere show attitude (or reaction control) thrusters that guide – via computer control – the vehicle into dock at the Space Station or final landing on the moon's surface.

The large circular depression just north of the cockpit (facing upward) contains the FLIR tracking optics and cameras. The central processing computer (CPU) converts the data into a two dimensional real-time line graphic in the cockpuit monitor.

Heat shields on tracks slide over the cockpit windows when leaving or re-entering Earth's atmosphere.

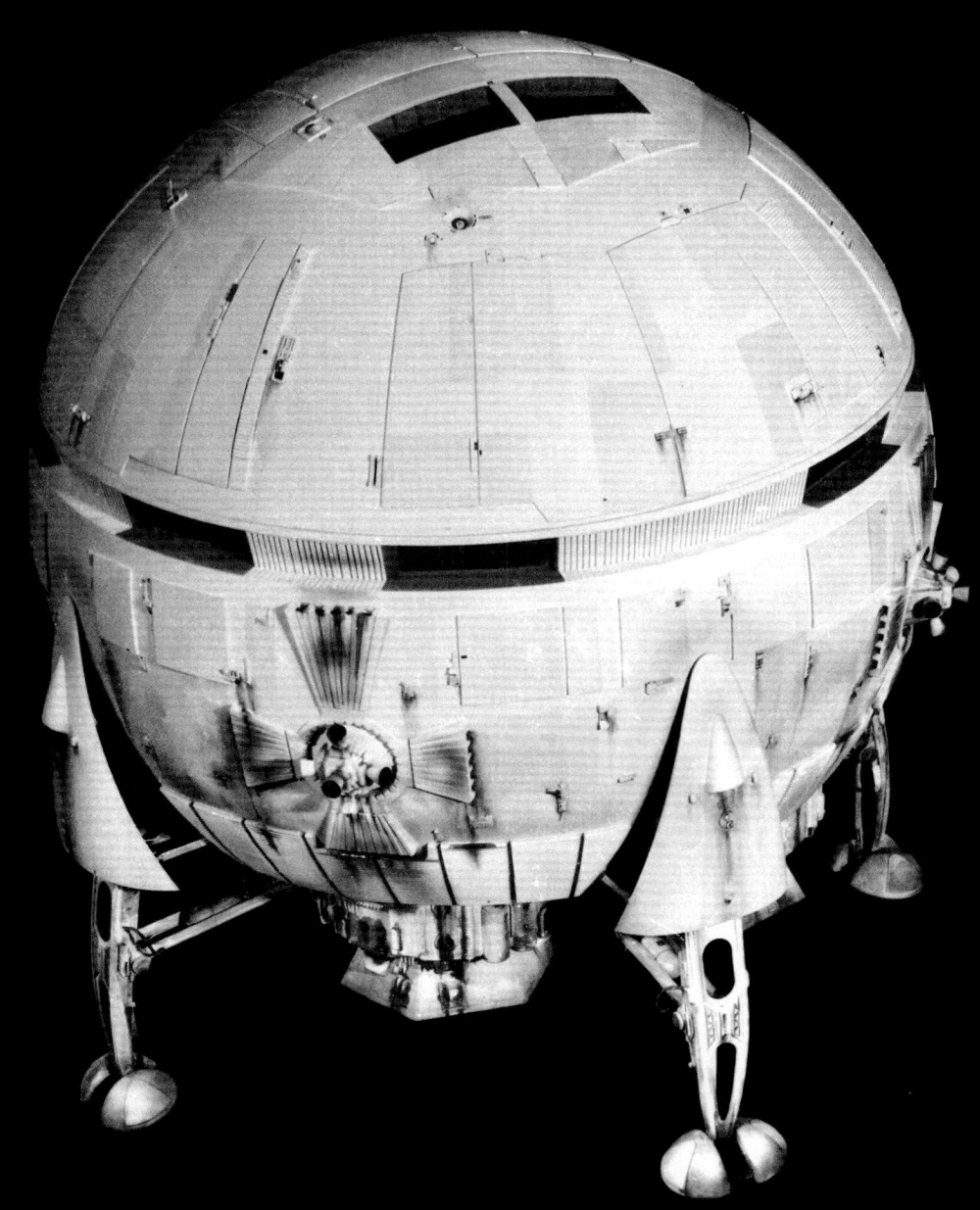

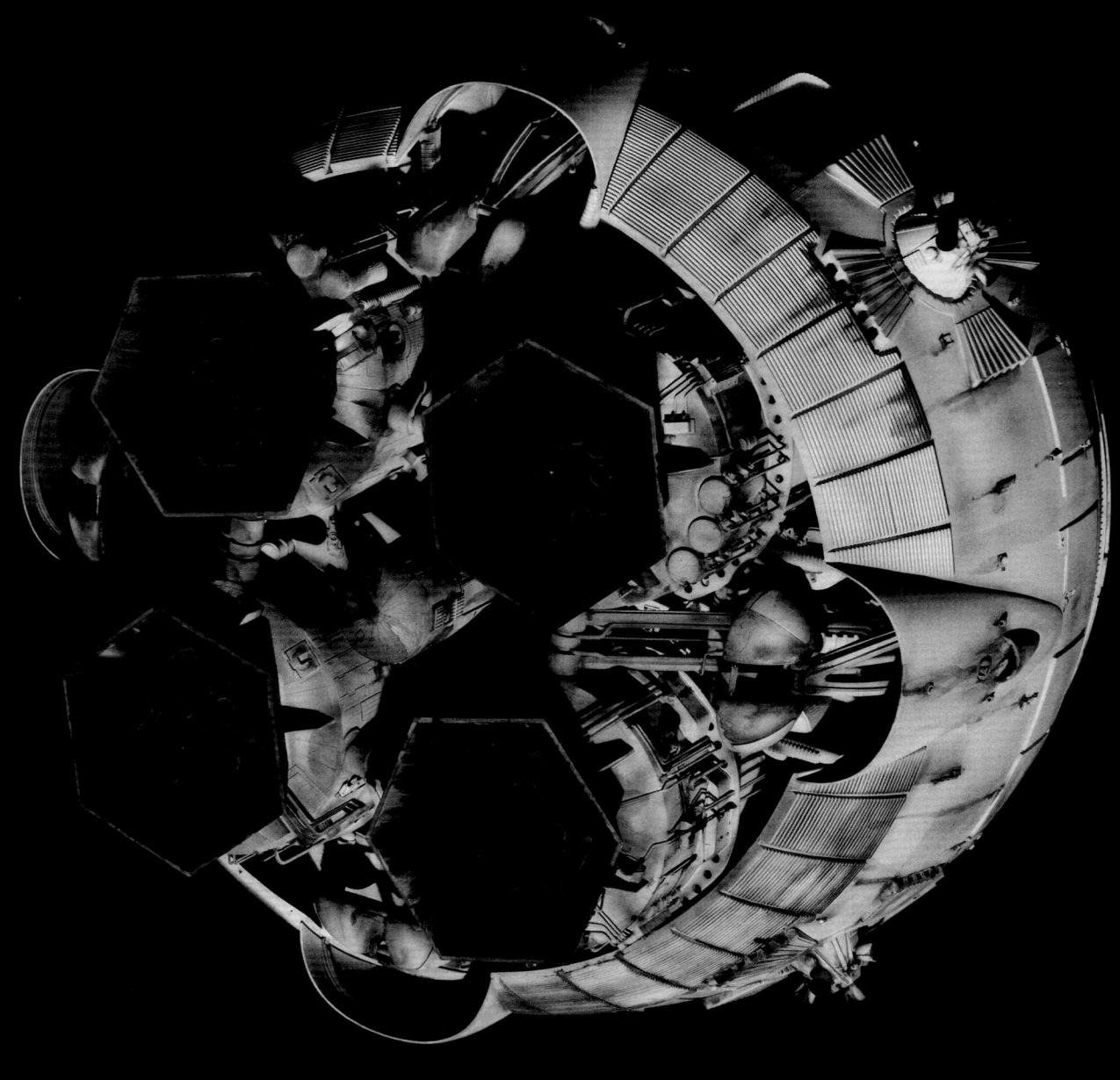

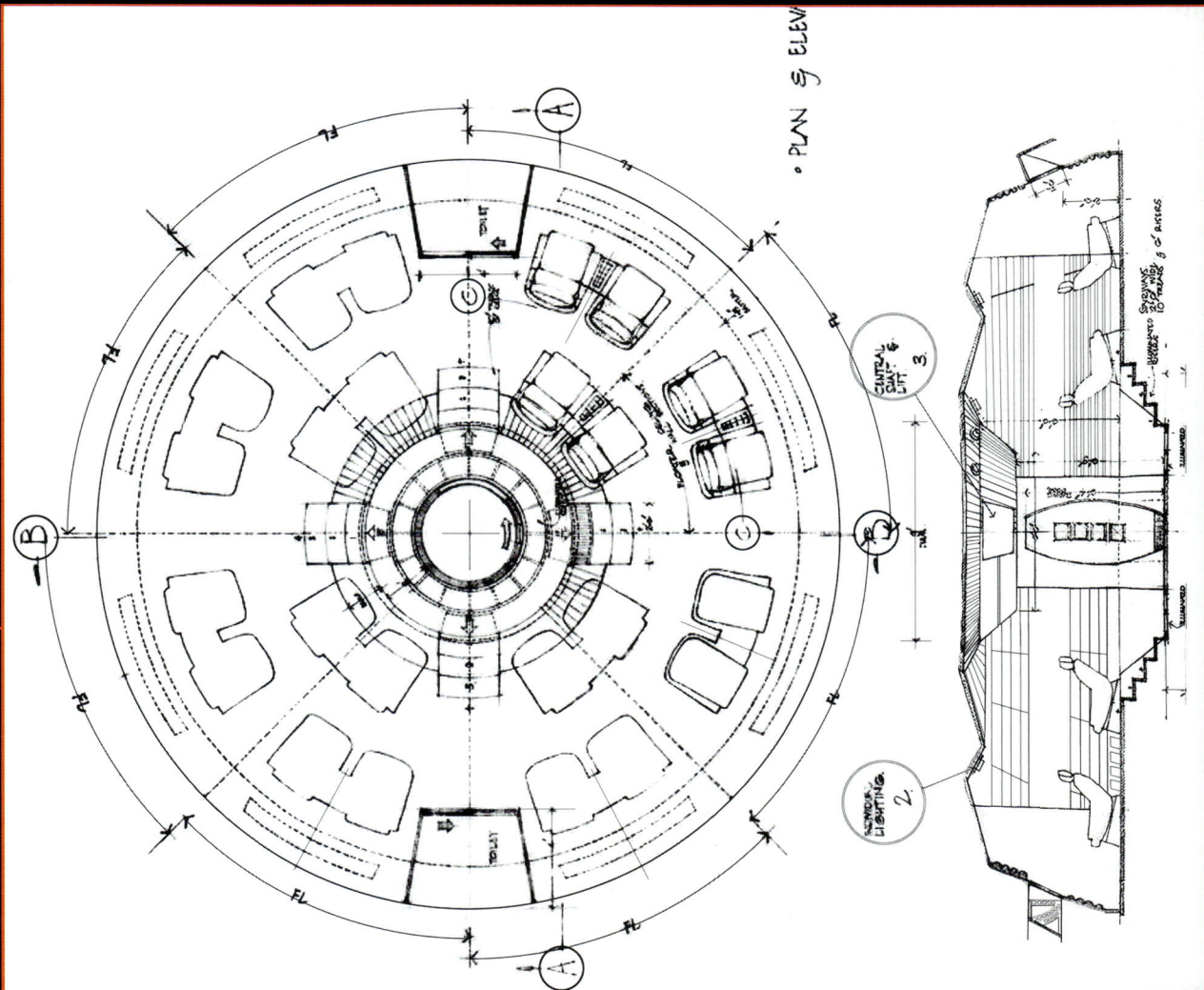

Aries 1B Passenger Deck Plan and Elevation

Views of the Passenger area show lavish accommodations a nd four large flat-panel widescreen televisions.

The Hawker Siddeley consultants on the Passenger Desk with Fred Ordway.

The Galley: Designed and Supplied by Whirlpool with electronics by RCA, provides many food choices for the discerning traveller. Due to zero-G conditions, food must be in a cream-like state and consumed through straws.

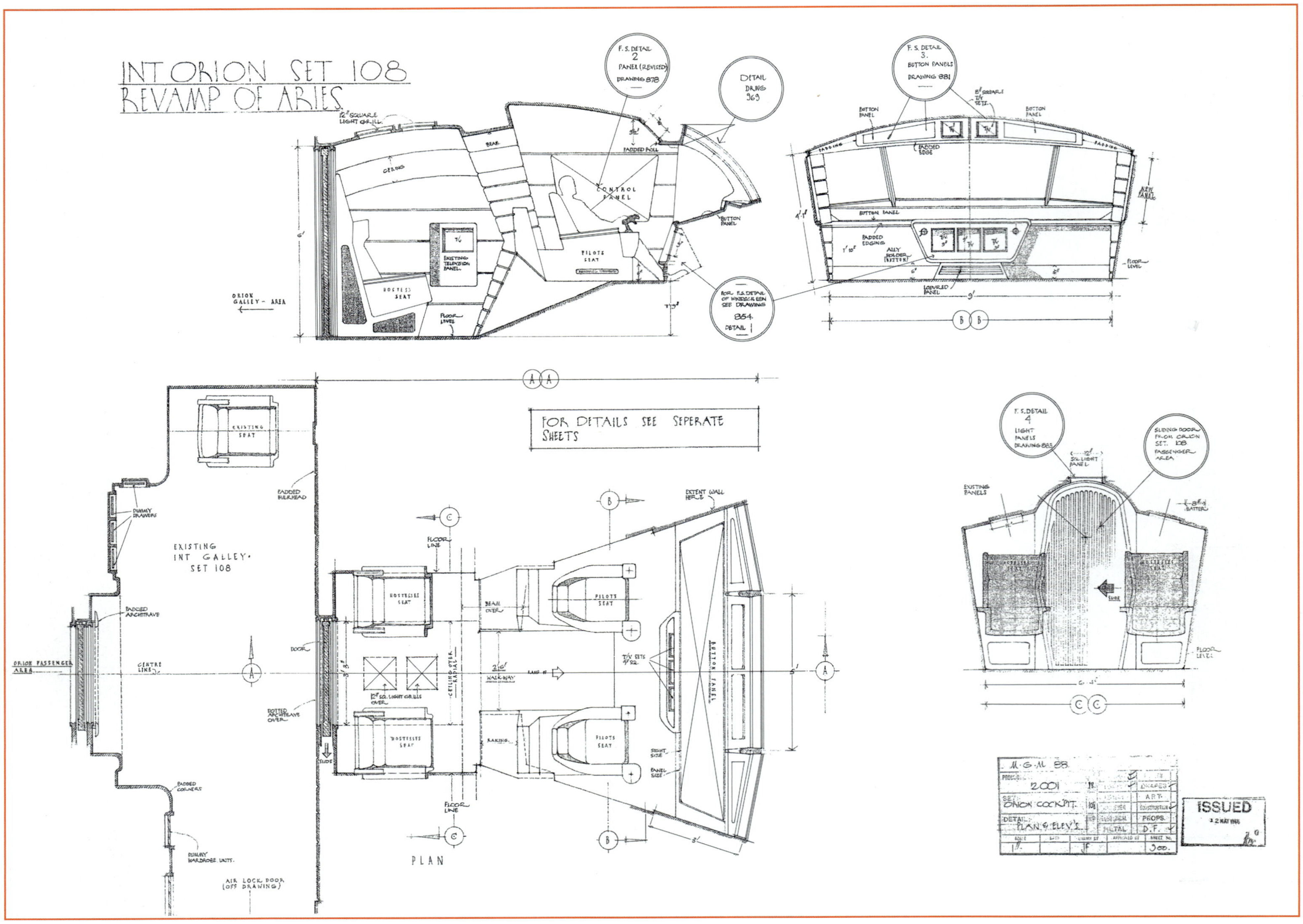

This schematic is for BOTH the Orion and Aries cockpits. The differences (see picture below) are a split window at the front of the Aries, a single computer monitor in the middle, and the elimination of the rear galley. The bulkhead behind the "hostess seats" encloses a circular elevator (not drawn) to allow movement between the galley at the lowest level, the Passenger Area, and the cockpit at the uppermost level.

Earth (Space Station) Time

Moonbase Time

Elapsed Flight Time

10 'level indicators' actively monitor cabin pressure, O_2 levels, temperature and various other routine functions.

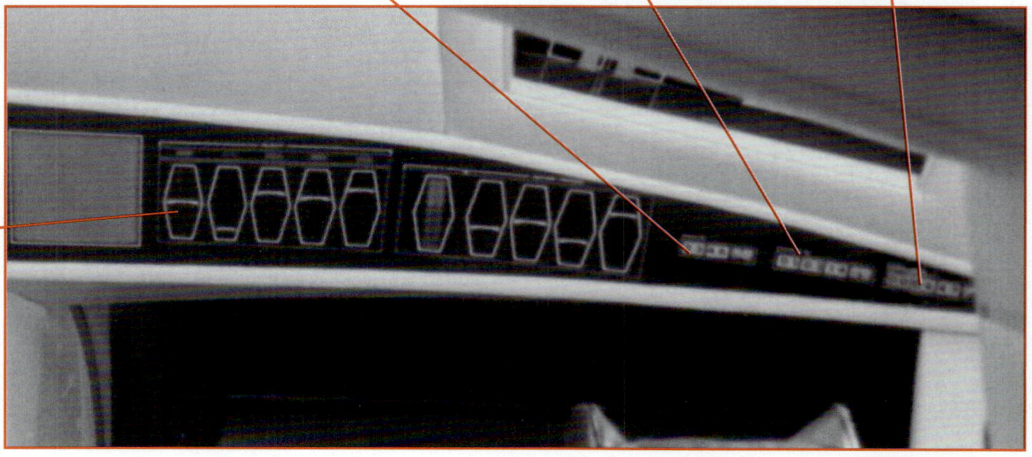

The IBM staff Ernest Bevilacqua, Gil Fox, Bob D'Arcona with Ordway and Lange during construction of the galley of the Aries. These people consulted on the computer systems onboard.

Ordway demonstrates how to use the seat-mounted controllers.

View near the Captain's command chair. To the left is a panel that displays detailed feedback and monitoring of the engines. The 'Dehavilland Vampire' joystick by his left hand guides the Aries. The 'hand' shaped controller by his right hand does critical approach and landing control via computer assistance, visually displayed by the main monitor between the two front windows.

This close-up of the four 'engine status' displays indicates real-time information, and the pilots can manually 'act' or 'de-act' any engine or group of engines. O2 and H2 fuel levels are shown by the long horizontal displays by the pilots' left upper arm.

31

CLAVIUS MOONBASE

The moonbase is located inside Clavius crater on the 'nearside' of the moon. It is almost a half-mile wide, with about 400 personnel stationed inside. Due to low gravity, people are cycled every 90 to 180 days. Designed and used specifically for scientific purposes, it also houses many families from allied nations who have voluntarily decided to live and work there. The moonbase is, in essence, a city. Curve walled structures are implemented for structural integrity to account for the dramatic pressure difference between the inside Earth-type environment and outside vacuum of space.

The architecture of the city is reminiscent of Ancient Rome, with the same circular geometric grid pattern and curved fortress-style walls. The key design feature here is the ever-expanding series of rings with a true central complex in which the community congregates. The individual structures are reminiscent of the style of architect Frank Lloyd Wright during the middle of the 20th century.

The entrance is a large underground hangar situated above, and the dome peels open. The Aries shuttle lands, then descends to a pressurized subterranean area. The dome is a protective cover against meteor or other bombardment by unusual solar system activity and the harsh solar winds and radiation. Magnetic rail shuttles move people and equipment via underground tunnels.

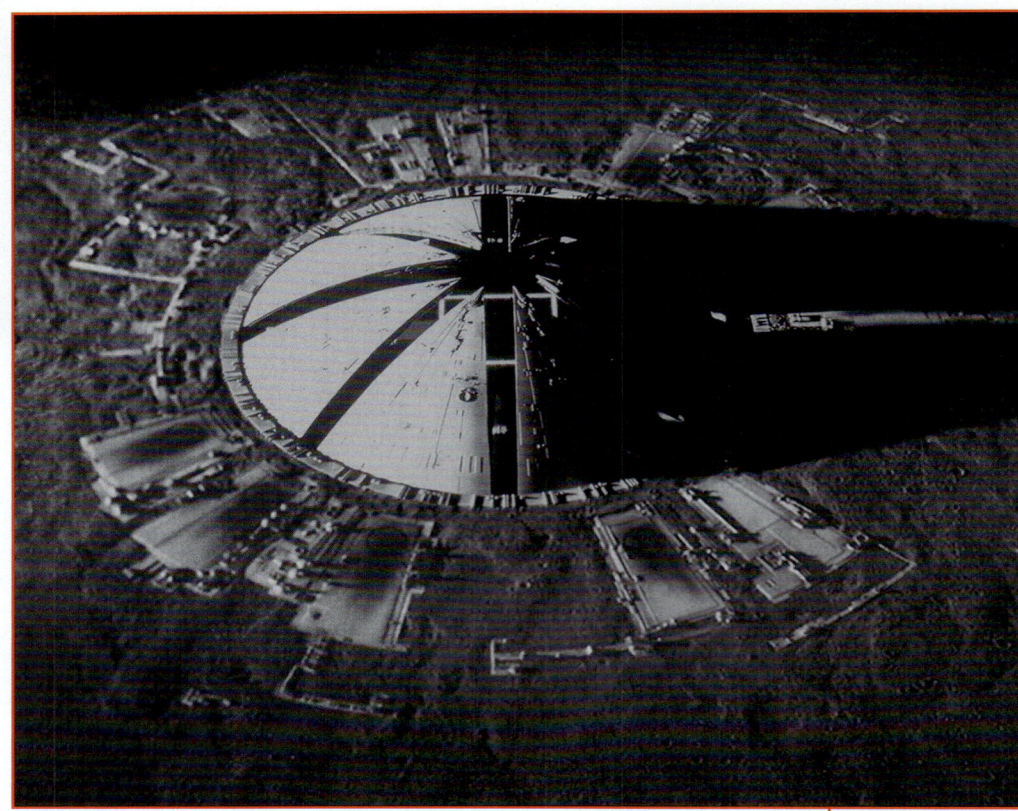

Three members of the United States Embassy being briefed at Clavius Base. Colonel James Michaelis, US Air attache; Dr. Boda Bartocha and Dr. Burt Edelson, Office of Naval Research. Such individuals provided key consultation in the function of the Moonbase.

Clavius Landing port protective dome

On approach to Clavius crater seen at an altitude of 400 meters.

The massive Clavius Moonbase City. The concentric rings of the city visually follows the craters of the moon. "Form follows function – that has been misunderstood. Form and function should be one, joined in a spiritual union." - Frank Lloyd Wright.

Left side of Moonbase City. Photo artificially changed perspective.   Right side of Moonbase City. Photo artificially changed perspective.

Frank Lloyd Wright's Guggenheim Museum in New York City. 1959.

Throughout the facility, photographers record the many high-level guests that visit the Moonbase every week. They use this Nikon camera. Its unique design makes it fast, user friendly, and has high speed auto-focus capability. Literally "point and shoot."

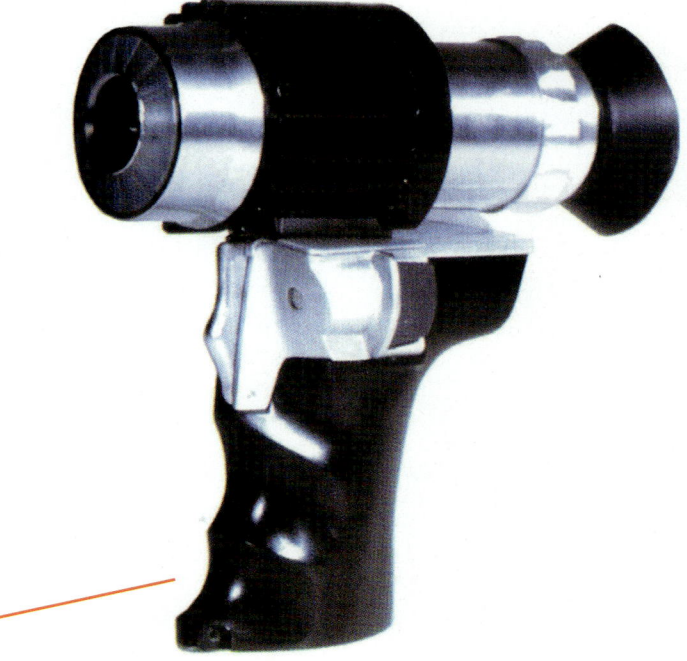

EXPLORATION OF THE MOON

The main mode of personnel and cargo transportation are Rocket Buses and Molabs across the lunar surface.

Early designs of the *Rocket Bus* originate from the 1965 NASA Molab vehicle (Mobile Laboratory) which shows similar design and functional properties, amongst them are the rear-entry door, and interior layout. Currently Molabs all across the moon move heavy equipment and personnel in and around Clavius Base and the various depots and excavation sites.

Mobile Geologic Laboratory designed for NASA by General Motors in 1965 as an extended lunar mission vehicle.

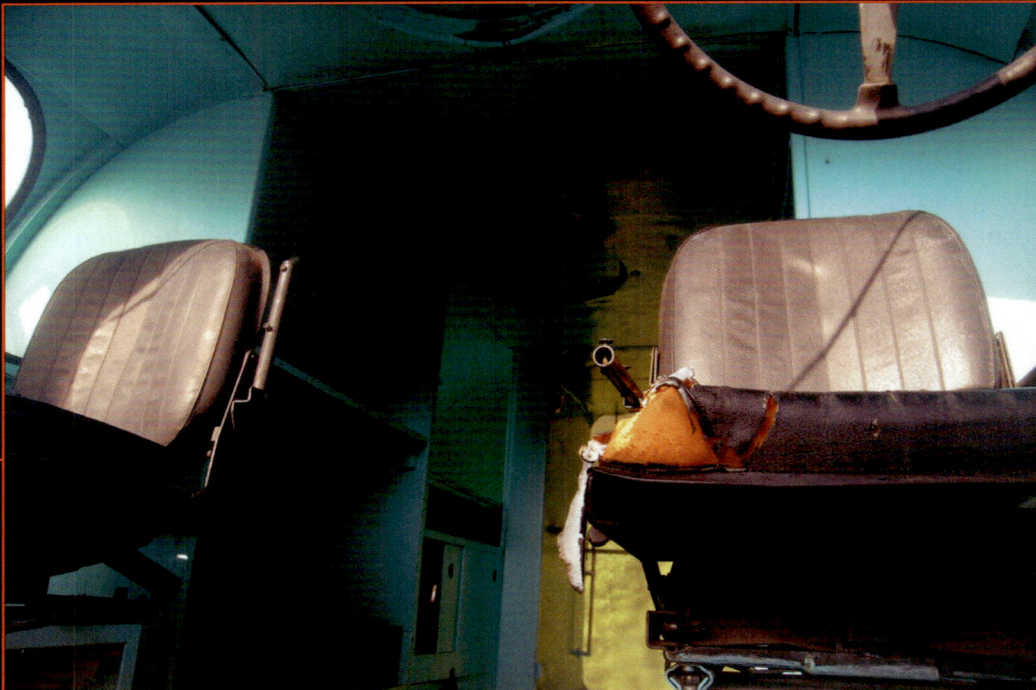

The interior of the prototype MOLAB as seen today, is almost identical in every way to the 'Rocket Bus', other than being more compact. Entry to the vehicle is through a rear door.

MOLAB tracks are seen here at an 'Emergency Depot' by Tycho Crater. The Landing Pad at right is where a Rocket Bus lands.

Tycho Crater is one of the largest craters on the moon and can be seen from Earth. At 53 miles wide, it is a key area of study to scientists. A temporary moonbase, "Emergency Depot", comprised of two habitat pressure spheres, storage barrel containers, and numerous excavation equipment, is set up to study the Artifact discovered there. The TMA-1 excavation is a mere 100 yards away.

The 'Rocket Bus' is a 38-foot rocket-propelled transport used for base-to-base and scientific survey across the moon. It is a hovering craft with translation movement supplied by vertically oriented rocket motors. Vents at the rear of the craft propel it forward by diverting the energy horizontally in short bursts from the rocket motors in the aft section.

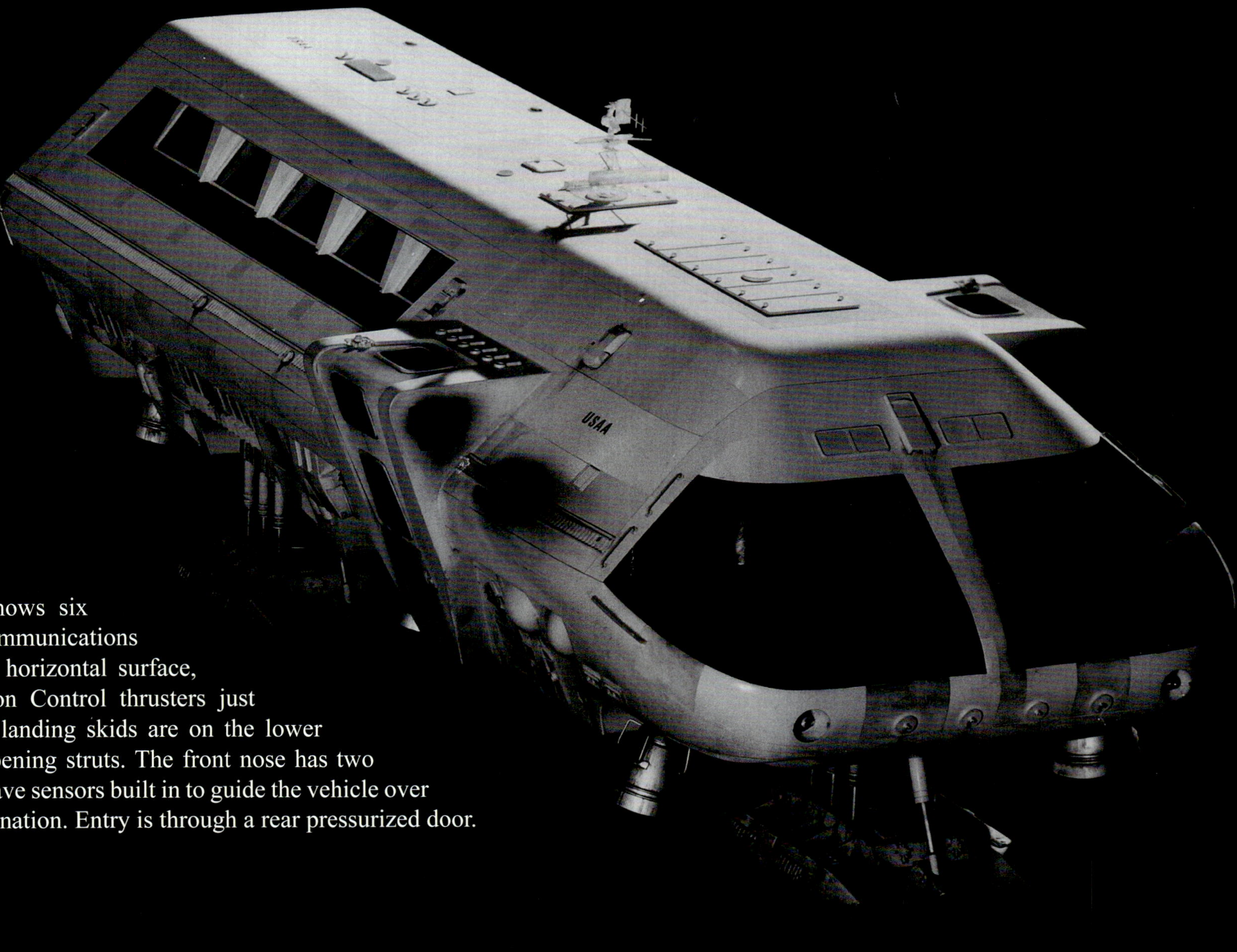

The 'Rocket Bus' here shows six translation thrusters, a communications antenna on the uppermost horizontal surface, and two banks of Reaction Control thrusters just behind the cockpit. Three landing skids are on the lower surface supported by dampening struts. The front nose has two lights and four nodes that have sensors built in to guide the vehicle over the lunar surface to its destination. Entry is through a rear pressurized door.

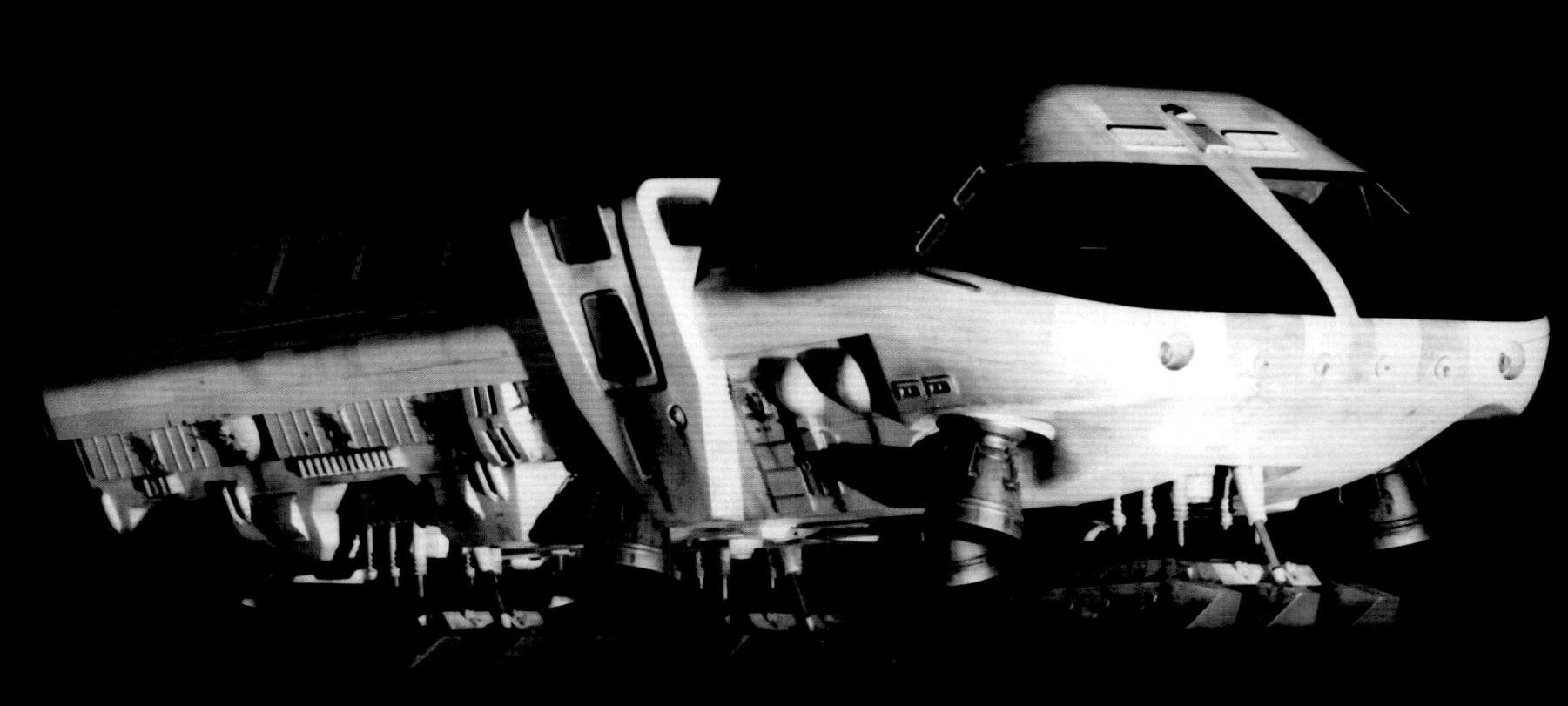

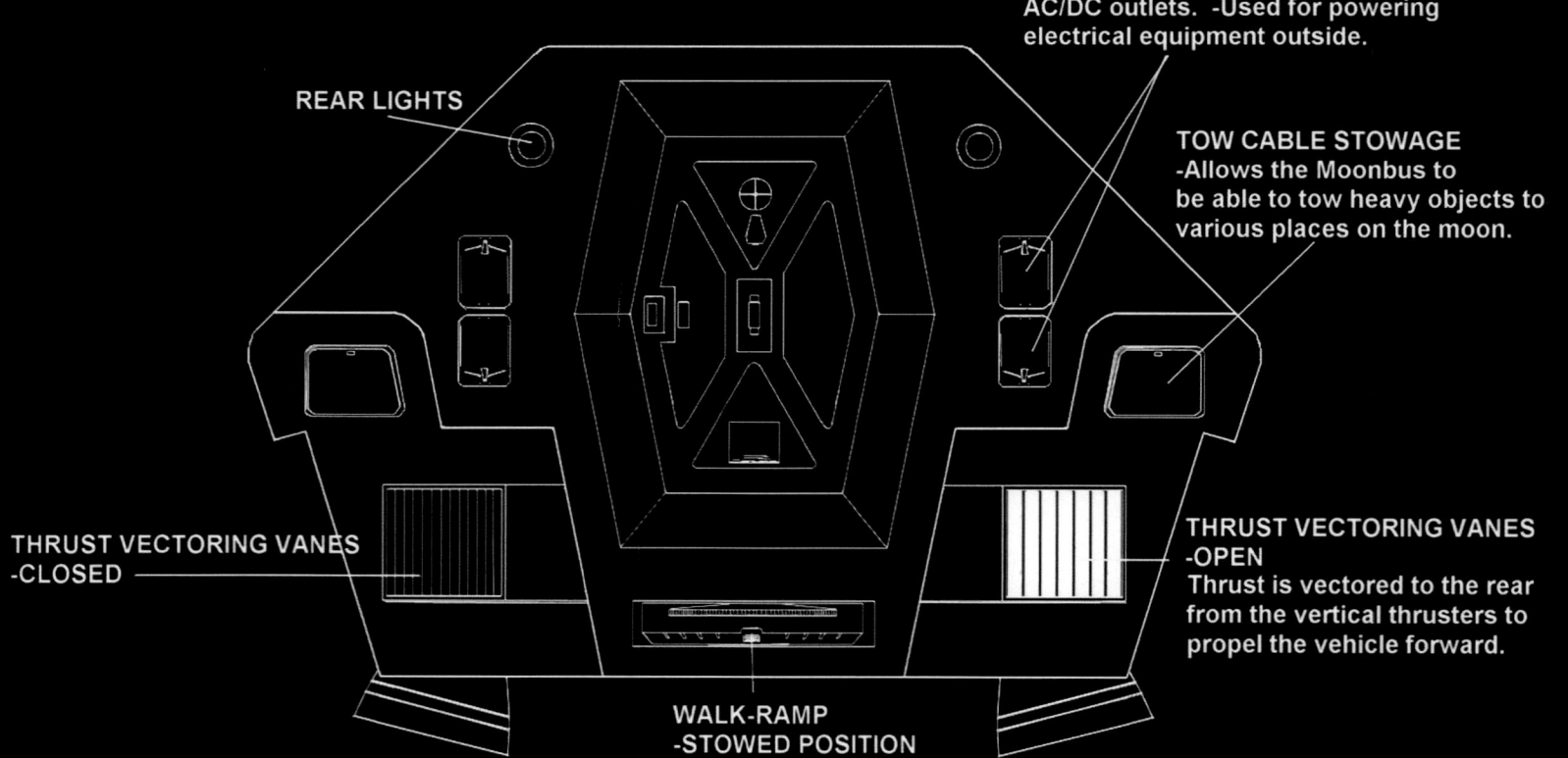

- **REAR LIGHTS**
- **AC/DC outlets.** -Used for powering electrical equipment outside.
- **TOW CABLE STOWAGE** -Allows the Moonbus to be able to tow heavy objects to various places on the moon.
- **THRUST VECTORING VANES -CLOSED**
- **THRUST VECTORING VANES -OPEN** Thrust is vectored to the rear from the vertical thrusters to propel the vehicle forward.
- **WALK-RAMP -STOWED POSITION**

A scientific team suits up and gets ready to embark outside the Rocket Bus. This is the area of the vehicle looking toward the rear entry/exit door.

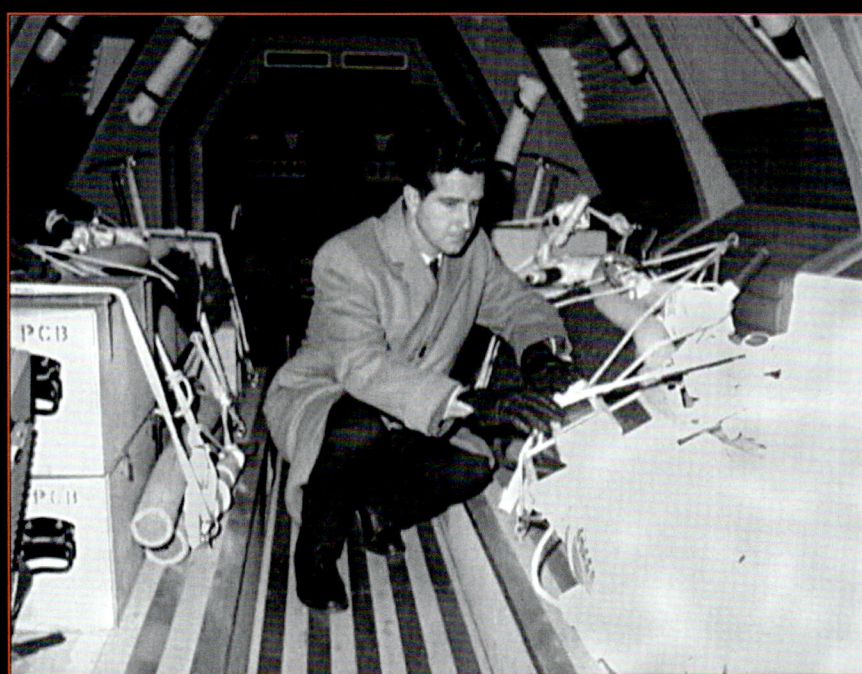

Ordway checks the tie-downs of the analyzing equipment in the Rocket Bus. Behind him is the cockpit. The equipment seen here is used for selenological studies, such as lunar magnetics, lunar geologic history, geophysical analysis and spectral analysis. The floor has magnetic strips to allow astronauts' boots to keep firmly attached to the floor in low-gravity conditions.

Fred Ordway stands in the mid-section of the Rocket Bus. This area stores supplies, and the complete cartographic maps of the moon. Also rock samples are gathered and contained in the lower bins.

The 'Rocket Bus' blueline drawing of the front cockpit horizontal control panel. (left side drawn; right side is a mirror image.)

A pre-schematic dimensional drawing of the cockpit control panel shows two sets of the 'hand-shaped' landing controllers. One set of them was removed, and the joystick moved in its place.

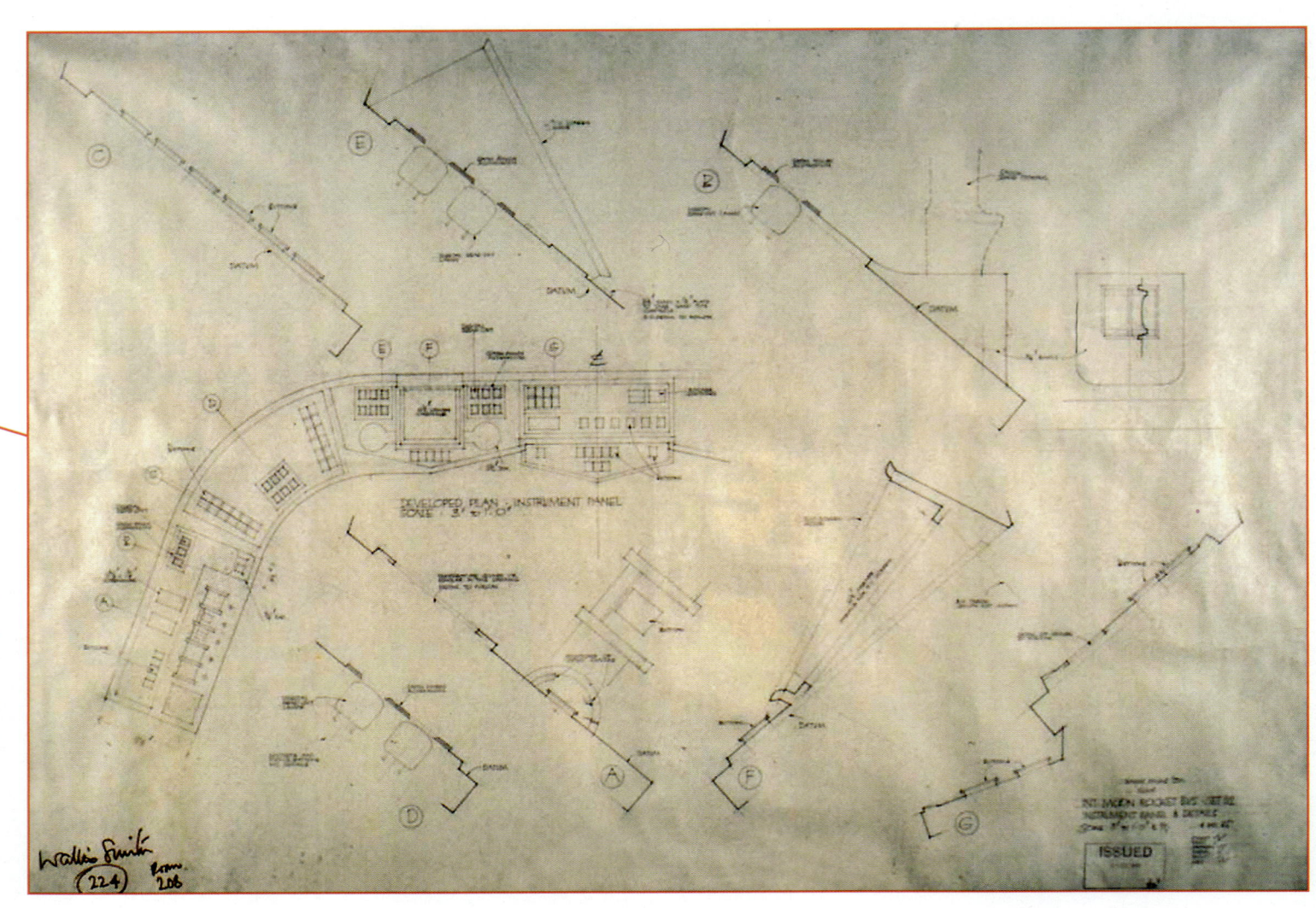

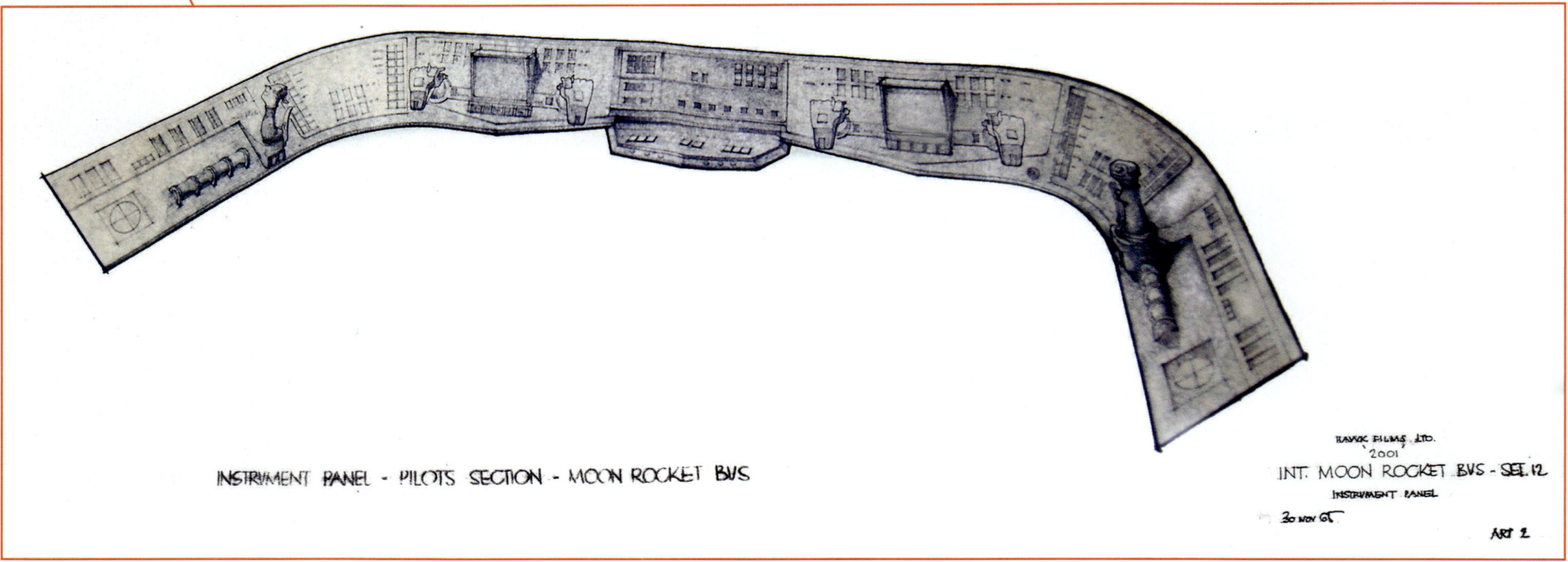

Ordway checks the functionality of all the controls in the cockpit. Note the high attention to ergonomics.

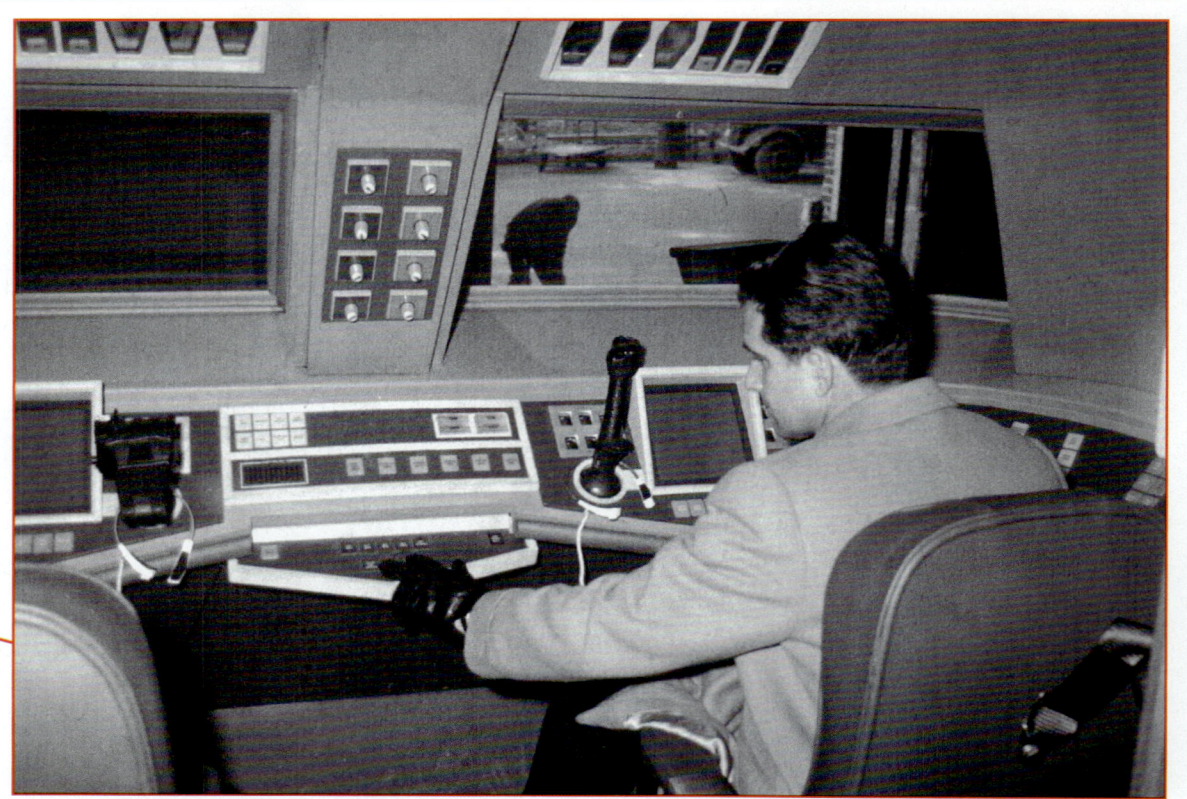

38

TYCHO MAGNETIC ANOMALY (TMA-1)

It is interesting to note that to an observer on the centre of the Moon's disk, the Earth appears overhead; but, as one goes north, south, east or west, it gradually sinks towards the horizon. Thus, Tycho's location at 43° south latitude, determines the Sun's angular height above the horizon. Moreover, since Tycho is 10° west longitude, the Sun is offset by that amount. The Earth is virtually fixed in the lunar sky, its slight movements only being noticed when it is near the horizon. All other bodies, such as stars, rise and set. A complete rotation of the lunar sky takes 29 days, so to observers at Tycho the stars slowly drift across the sky along with the Sun – which would take about an hour to completely rise or set.

At the TMA-1 site, the principal technical features were the array of instrumentation surrounding the newly excavated rectangular-shaped 10-ft high block which occupied the central theme in the film. The devices were designed both to determine any potential danger associated with the block and what its nature might be.

The radiation monitor measured the radioactivity level: the intensity of alpha and beta particles, gamma rays, and neutrons in the energy range from 0.1 Mev to 1000 Mev in units of counts per second or Rads, accomplished by setting the appropriate dials. Magnetic tests were conducted with a hydrogen precession magnetometer. A compact computer-oscilloscope was employed to manipulate, store and display the data, with other sensors measuring different physical properties. For example, a sound pulse transmitter-receiver, a radiation detector and infrared-image orthicon were all connected to the computer oscilloscope by cables. Each sensor could be monitored individually and its displays magnified, mathematically manipulated, etc.

The infrared-image orthicon was designed to transmit heat pictures of the structure. The sound-transducer was held against the block to produce short-duration sound pressure pulses. As these were propagated through the slab, echos would be reflected off the different internal structures, sensed by the transducer, and then displayed. The intensity of the returning echos would be shown as a function of the time-lag in milliseconds from the initial pulse. A complex inner structure would be indicated by complex echo patterns.

To attempt to determine what elements were in the block, high energy protons, alpha particles, beta particles, neutrons and gamma rays in the energy range from 0.5 to 100 Mev were scattered by the linear accelerator. It swept across this energy-range whichever type of radiation was selected, and the tripod-mounted radiation counter measured the reflected radiation for each energy. Information was transmitted back as intensity in counts per second versus energy. Sharp decreases in reflected radiation would indicate resonant absorption for that particular energy – a characteristic of each element. Many absorption peaks would have indicated a complex structure.

The Tycho construction site 'TMA-1.' Technicians complete the support walls and assembly lighting and other analyzing equipment for detailed analysis of the monolithic artifact that was buried several meters under the lunar surface. Against the rear wall is a portable laboratory. On the upper surface are solar panels to convert and store electricity to run the temporary facility. From here, scientists collect data.

A scientific team walks down to observe the monolith. Around the artifact are four devices that take detailed analyses. (next page for description).

Radioactivity Level Monitor

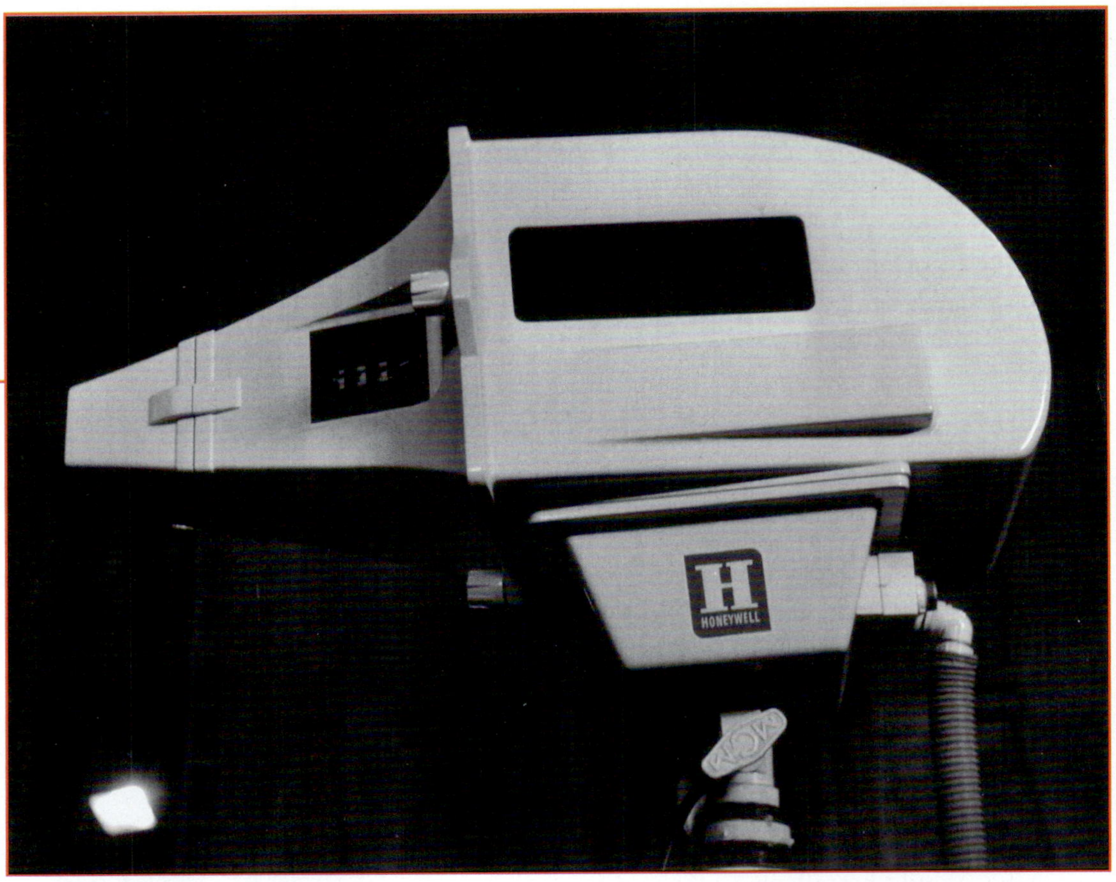

Infra-red Imaging Station

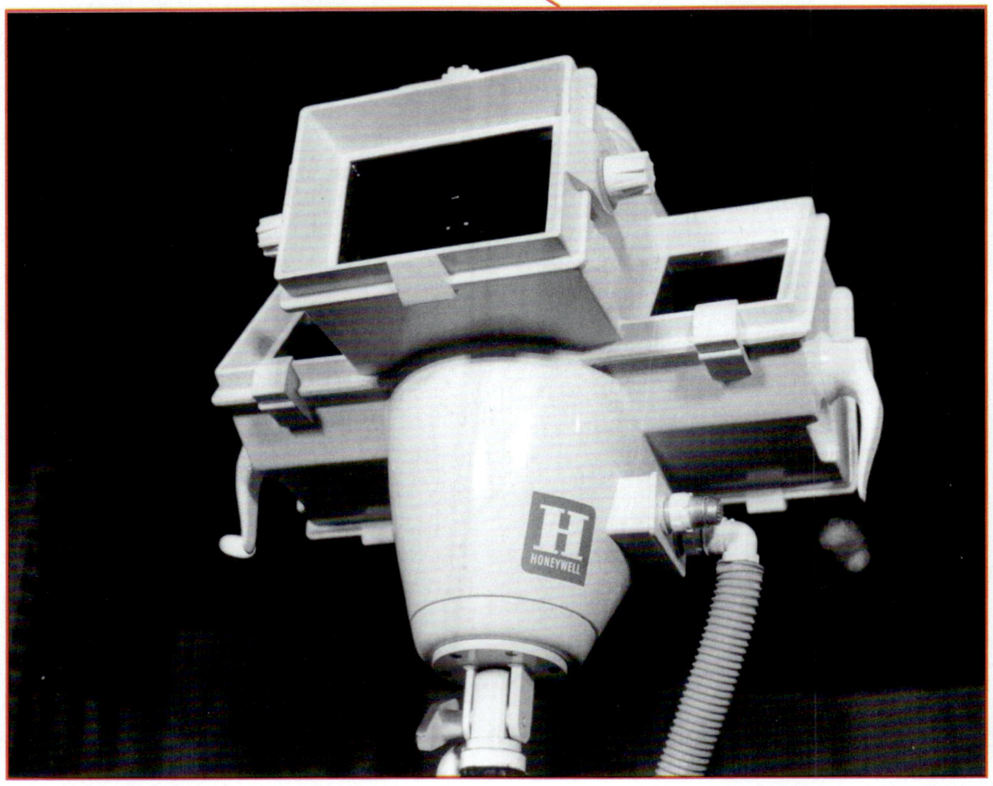

Particle Scattering Transmitter/Receiver

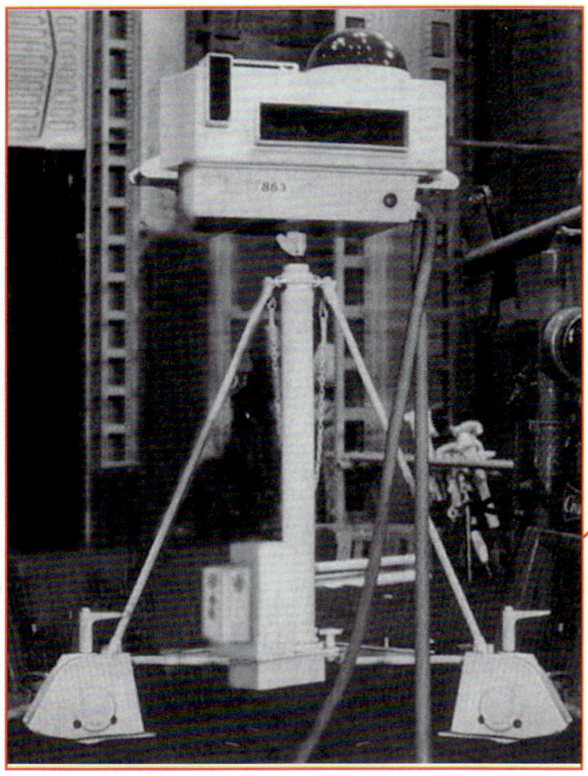

Magnetometer

40

THE DISCOVERY

Space Vehicle Design

The many space vehicle designs created for M-G-M were filmed between January, 1965 and the Spring of 1966. We had a support staff of some 35 designers, many from the British aerospace industry. In addition, we were helped by scores of outside engineers.

The concepts for the vehicles were dictated by our experience in the spaceflight and rocket field up to mid-1965 and extrapolations three decades into the future. We had previously been working on a joint Marshall Space Flight Center-General Electric payloads requirements study for the planned Saturn 5-N (for Nuclear) type vehicles designed for Mars missions. These studies called for the incorporation of a nuclear 3rd-stage designated S-NB. As a 'suborbit start stage,' it was to have used a Mod 2 Nerva engine. Initial work involved statements of mission objectives, modes of operation, parametric analyses, special mission and design problems, evaluation of systems performance, cost, trade-off sensitivities, definition of ground support concepts, etc.

Studies in 1964-65 showed that by clustering nuclear stages of modules in orbit, it would be possible to achieve manned orbiting, and possibly landing, flights to Mars within the next 15 years. One concept derived from Empire* and Post-Empire* studies, called for each primary propulsion module to be powered by a 195,000-lb thrust Nerva engine. The propellant tank would be 33 ft in diameter, 115 ft long, with a capacity of 385,000 lbs of propellant. Each module would be fitted with an 8-in diameter line to transfer propellants, as necessary, from one module to another during an interplanetary mission. Overall space vehicle length was 108 ft, consisting of the propulsion modules, the main mission module, aft interstage compartment, airlock, and an excursion module which would descend to the surface of Mars after the main spaceship had entered orbit.

The Discovery

The vehicle selected for the Jupiter mission was the impressive 700-ft long spaceship *Discovery*. Gaseous core nuclear-reactor engines at the rear provide the craft's propulsion. Hundreds of feet of tankage and structure separate this from the spherical part of the ship where the crew quarters, the computer, flight controls, small auxiliary craft, and instrumentation are located. In the *centrifuge*, the crew enjoy Earth-like gravity conditions created by spinning; it is there they spend most of their time and where the hibernating astronauts sleep in their 'hibernacula.' Actual piloting, navigational checks, and the like take place in the zero-gravity environment *command module*. Other sections of the sphere include the *pod bay*, where three one-man repair and inspection craft are housed, and the HAL 9000 computer 'brain room' with its level-upon-level of memory-storage and related elements reside.

In developing our design concept for *Discovery*, we used many available published reports and gradually evolved the vehicle which appears in the film.

The Command Module, from where the *Discovery* is piloted during take-off from Earth orbit, midcourse corrections, and Jovian capture maneuvers has a broad window visibility. During the long, interplanetary coast period this area is not normally used by the crew, who remain in the Centrifuge where they enjoy artificial gravity.

Propulsion controls are located in the command module. The nuclear reactor control panel displays information on such parameters as: turbine, compressor, heat exchanger, secondary circulatory and radiator liquid, helium storage, MHD generator and recuperator performance, and pressures and temperatures at various stations. Precise present readings can be obtained instantaneously on the control screen, if desired, as well as past performance and predicted future performance.

The Cavradyne engines are based on the assumption of years of research and development during the 1980s and 90s, of gaseous core nuclear reactors and high temperature ionized gases or plasmas. Theory is presumed to have shown that gaseous uranium 235 could be made critical in a cavity reactor only several feet in diameter if the uranium atomic density were kept high, and if temperatures were maintained at a minimum of 20,000ºF. At first, progress was slow because of such early unsolved problems as how to reduce vortex turbulence in order to achieve high separation ratios, and how to achieve adequate wall cooling in the face of the thermal radiation from the high temperature ionized plasma. In the Cavradyne system, the temperature of the reactor is not directly limited by the capabilities of solid materials, since the central cavity is surrounded by a thick graphite wall that 'moderates' the neutrons, reflecting most of them back into the cavity. Wall cooling is assured by circulating the hydrogen propellant prior to its being heated. Fissionable fuel energy is transferred to the propellant by radiation through a specially designed rigid -- and coolable – container.

*Empire: Early manned planetary-interplanetary roundtrip expeditions.

This paper summarizes the EMPIRE studies undertaken by the NASA-Marshall Space Flight Center and its Aeronutronic and General Dynamics/Astronautics contractors in the early 1960s. Among topics addressed are duel-planet (Mars and Venus) trajectory studies and mission requirement, guidance and navigation, space vehicle configuration, propulsion, crew and environmental considerations, science payloads, scheduling, and funding. Units employed in the source literature are cited throughout.

See accompanying DVD for complete Empire studies abstract by Fred I. Ordway III.

*Post-Empire studies:

Manned Mars Exploration in the unfavorable time period (1975-1985), by General Dynamics and Douglas Missiles and Space Division.

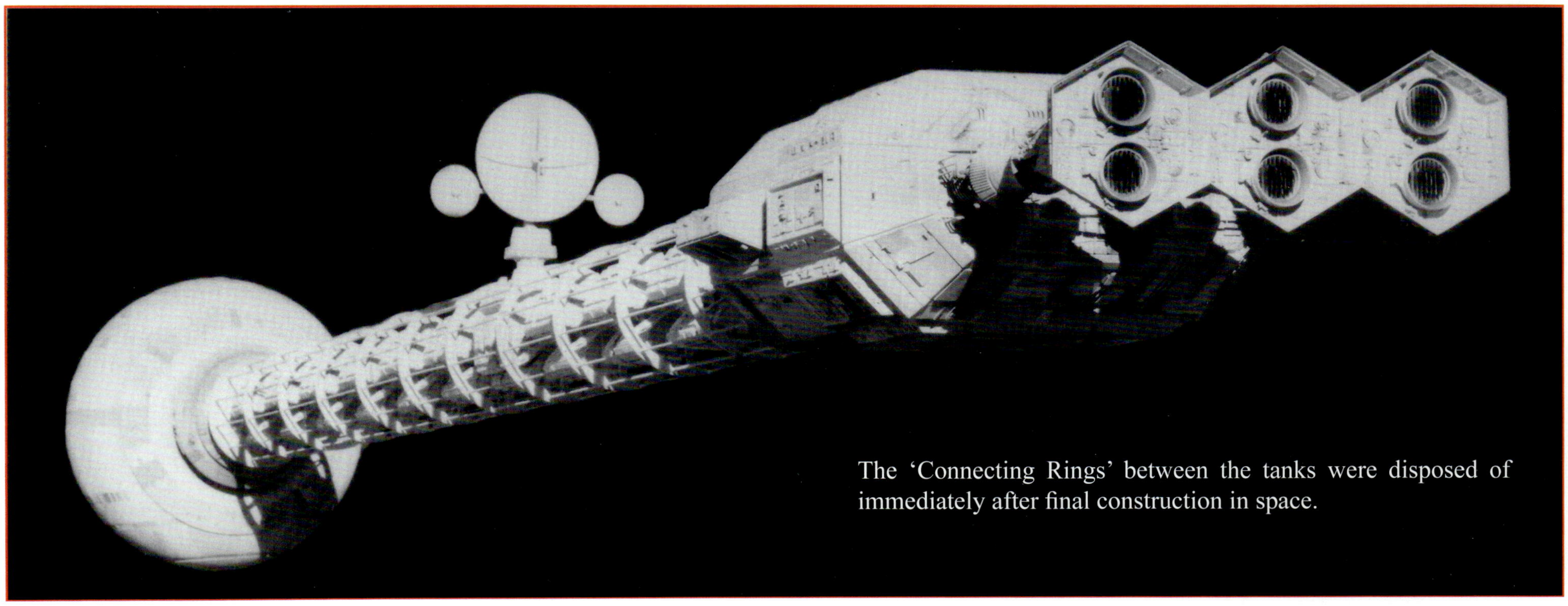

The 'Connecting Rings' between the tanks were disposed of immediately after final construction in space.

In this letter from Kubrick to Ordway, many detailed scientific extrapolations and designs came forth. Ordway consulted on a book by Samuel Glasstone to help with the new propulsion system on *Discovery*.

Samuel Glasstone authored 40 popular textbooks on physical chemistry, reaction rates, nuclear weapons effects, nuclear reactor engineering, Mars, space sciences, the environmental effects of nuclear energy and nuclear testing. The book, *The Sourcebook on Atomic Energy*, published in 1950, is still considered one of the best reference textbooks on the subject.

Discovery

MEMORANDUM

To: Fred Ordway
From: Stanley Kubick
16th Sept. 1965

"2 0 0 1"

Dear Fred:

Could you please work out a brief, concise explanation of the propulsion system and general operating features of Discovery? This will be used for Bowman explaining to a Television Interviewer something about the ship.

I am also still awaiting your rough breakdown of acceleration; distance travelled; velocity, and whatever perimeters might be interesting for the basic phases of the mission: - First day, first week, middle of the mission, deceleration, etc.,

Stanley Kubrick

(Dictated by Mr. Kubrick and signed in his absence)

Marshall Space Flight Center   Future Projects Office

Meeting on   March with Harry O Ruppe, director

Discussion of monitors on board Jovian spaceship:

(1) Propulsion

(2) Life support--suggested that we figure "reserve days" via computer--if negative, in terms of total mission days versus daily usage, then either usage rate must be lowered or mission speeded up. If a leak has been detected, computer would learn, signal warning, not merely increase air production to compensate for failure.

(3) Scientific equipment

(4) Probes sent into Jovian atmosphere--should be a semiautomated checkout procedure; review those used on Mariner flights

(5) Nuclear reactor--must write for data on how they are checked out or check in Glasstone or Lockheed sources

A self-controlling system should be available. A System to check out the checkout system; if the check response is OK, then the system is OK. To repair, have spare card modules for the computer--perhaps ten of each type.

Navigation: use a telescope to photograph the planets and the Sun against the star background--take photo to a photocomparator and get the accuracy of a fraction of second of arc. For the Sun take photos 180 degrees apart thus:

Information on the apparent diameter of the Sun is also good.

Another scheme is illustrated at right:

direct the speciman panel along the Sun-line; a radiometer measures the radiated temperature. Emission of the piene is known in terms of amount of light received by it. Emission is measured, therefore, distance from the Sun can be readily computer.

$$\varepsilon T^4 = \frac{\alpha}{L^2}$$

The computer aboard can perform a household function. Use, for example, strain gages throughout the vehicle's structure. Data fed to computer. If structure at one place shows signs of fatigue, the computer would flash a "stop centrifuge spinning" notice. If all is within tolerences, there would be no display. Monitors could monitor the oxygen partial pressure....if it goes outside limits, bell could ring and visual readout displayed.

There should also be a command override. Press button because of computer malfunction and no readout results. EMERGENCY

Clock could go out of operation. So, astronomical observations of the Jovian moons could be made. Depending on just what their instantaneous position is, the reading could be compared with the tables of published epherimedes, in which positions at all Julien dates and times appear.

The Alpha-Echo 35 Antenna Array

The large central dish is used for Audio/Visual communications with Earth. The smaller two are two-dimensional Aperture Antennas that link with the Discovery's HAL 9000 computer and determine Universal time, Sidereal time, distance to earth, distance to Jupiter, and the vehicle's general position in space. This is monitored at the communications module in the centrifuge.

Exploration of Jupiter

Once Jupiter had been reached and *Discovery* safely placed in elliptical orbit around it, the question arose as to what it would do after initial exploration of the planet and its moon had been made. The difficult situation of possibly having to go into orbit around the nearest moon, Jupiter 5, was investigated. Because of the moon's low gravitational field, the orbit velocity of Discovery would be very slow and station keeping would be required due to perturbations by the primary and the large Galilean satellites. In fact, a permanent orbit around Jupiter 5 seemed virtually impossible. If a low altitude were selected, inhomogeneities in its internal structure would probably result in a wobbly orbit which, over a period of time, might result in such elongation that the perigee would intersect the surface. If the orbit were too high, say one Jupiter 5 radius, then perturbations from Jupiter and its other moons would so disturb the orbit that eventually *Discovery* would be ejected. An orbital altitude of 3 radii gave us a period of approximately 4½ hours. It was estimated that a force of only 1/1000 g would eject *Discovery* from this orbit. This was an important consideration, since the mission plan called for all astronauts aboard to go into hibernation for perhaps 5 to 10 years after they had performed their exploration and before they could be rescued from Earth, and we had to find out where we should and where we should *not* leave the spaceship.

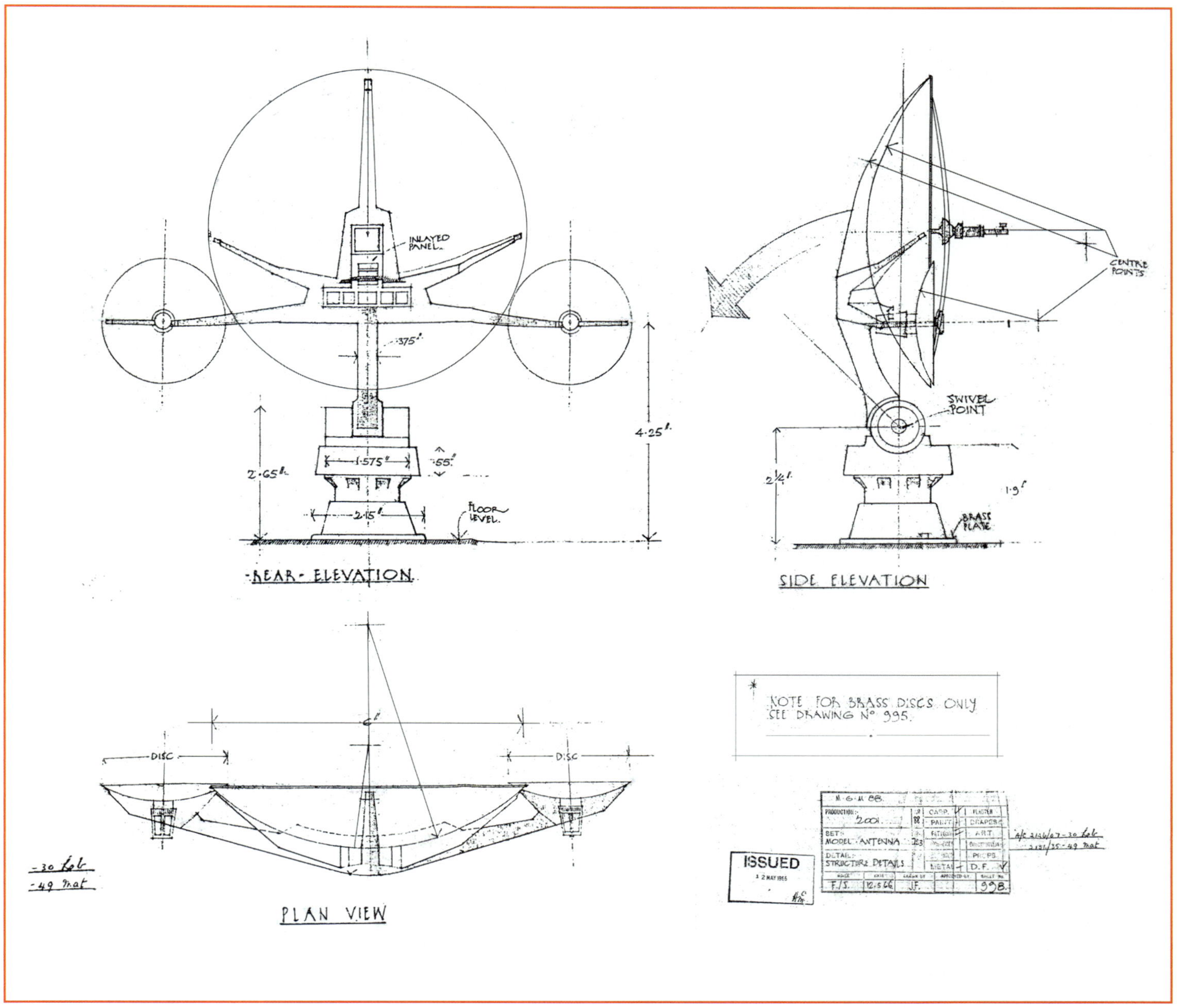

The Alpha-Echo 35 Antenna Array being tested prior to fitting in space. Main Dish is 14 ft across.

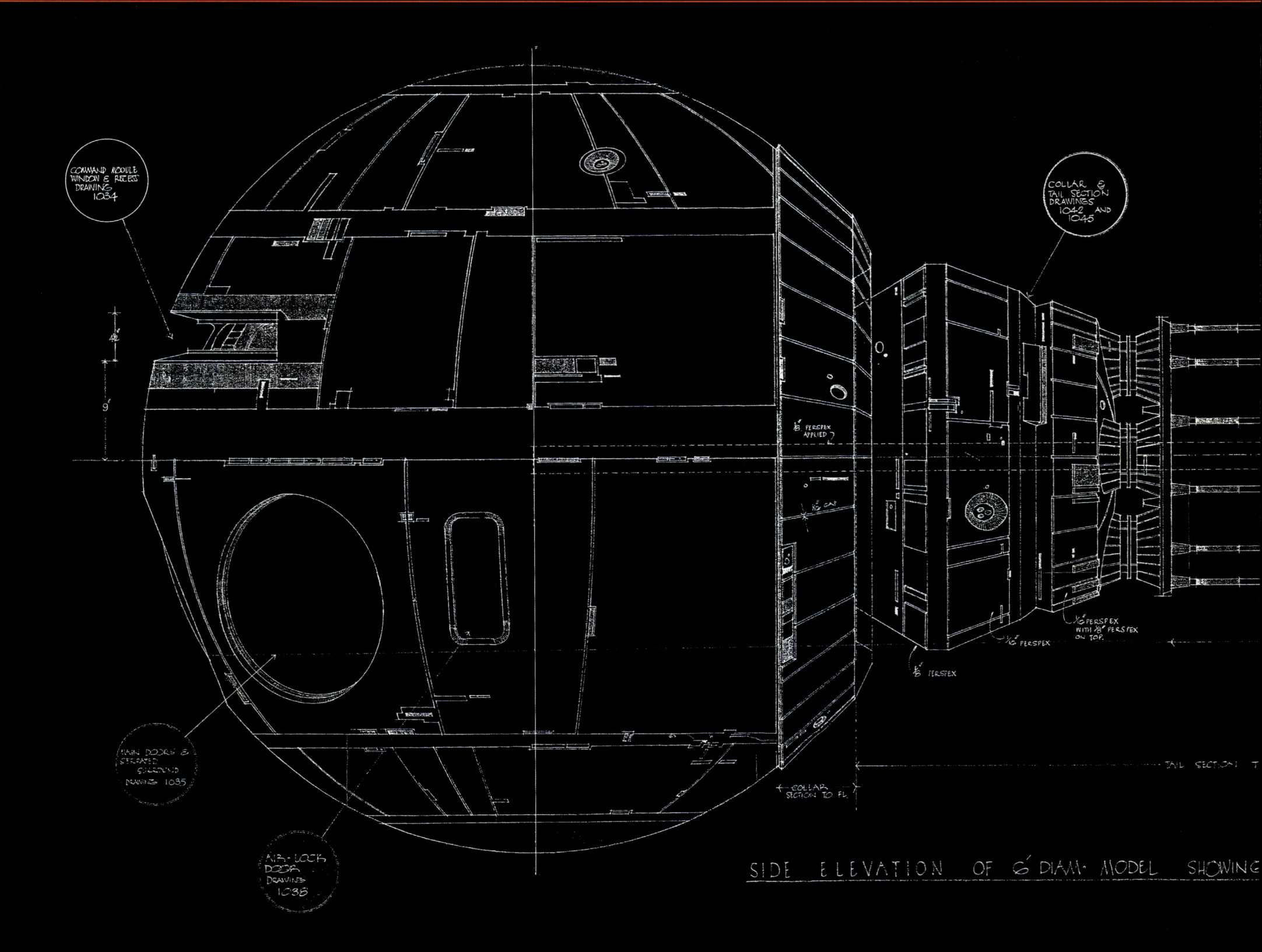

The Master Drawing of the *Discovery*. The front sphere contains all the habitat areas, at the rear is four chemical rockets that disconnect from the 'Tankage Structure' in an emergency or upon re-entering Earth's atmosphere. Note: 'Main doors' and 'Airlock' are overscaled on this drawing. Details can be seen later in this chapter.

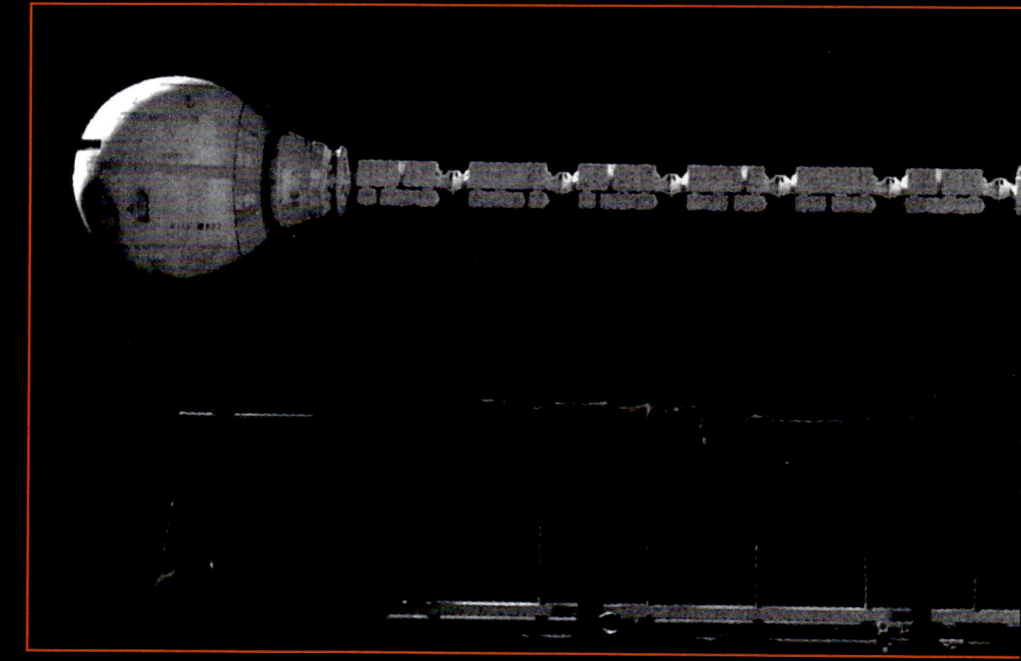

The Discovery is 520 feet long with a Habitat sphere 52 feet in diameter. The mid-section are storage tanks of waste and renewables during the 3 year trip to Jupiter.

Manual 'Habitat' overide release point.

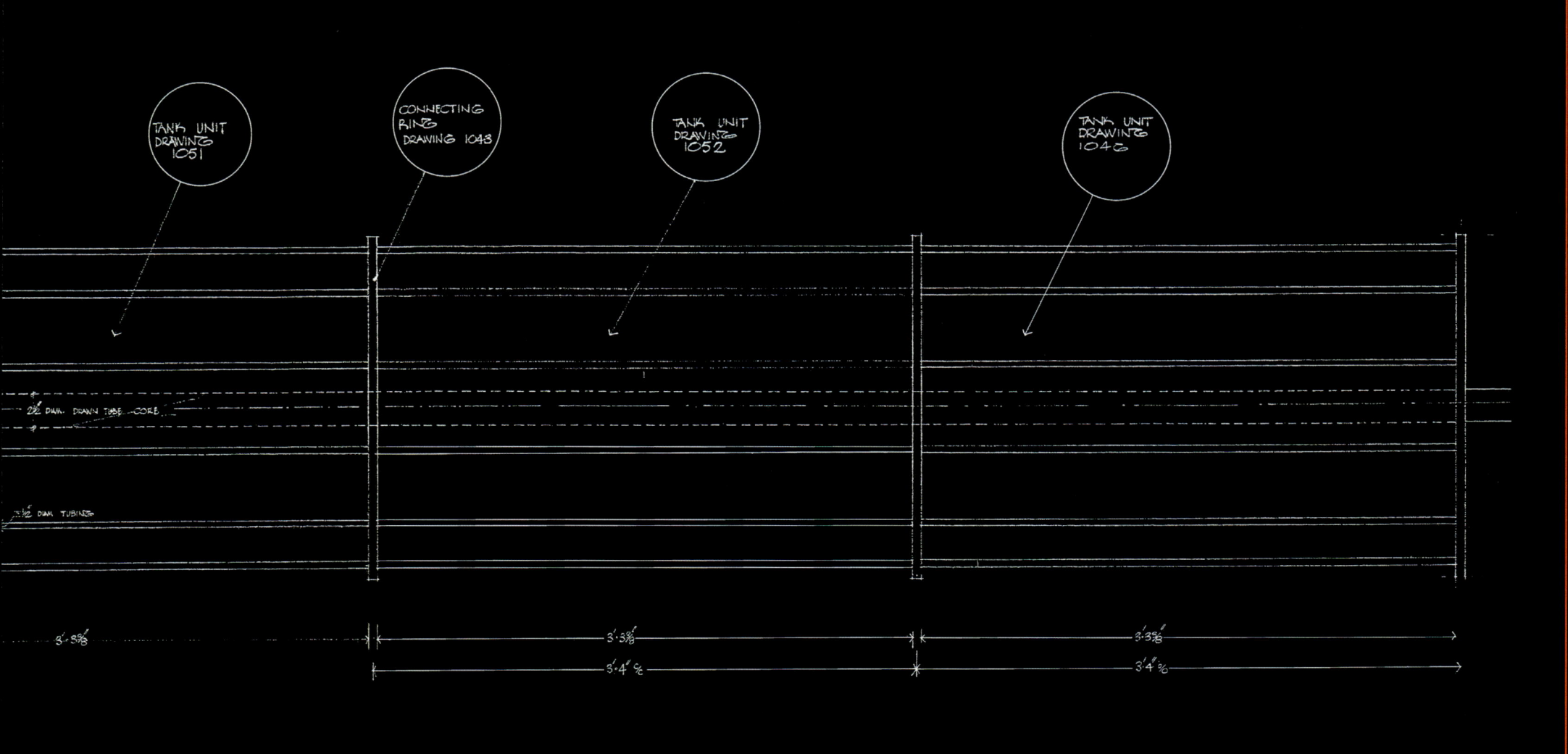

The Habitat sphere and collar can be removed from tankage via four chemical thrusters at the back.

Mid-section shows the communications antenna, and rear is the Cavradyne Reactor engines. 'Collar' section has 8 small round vernier thrusters used for precision control of attitude in space.

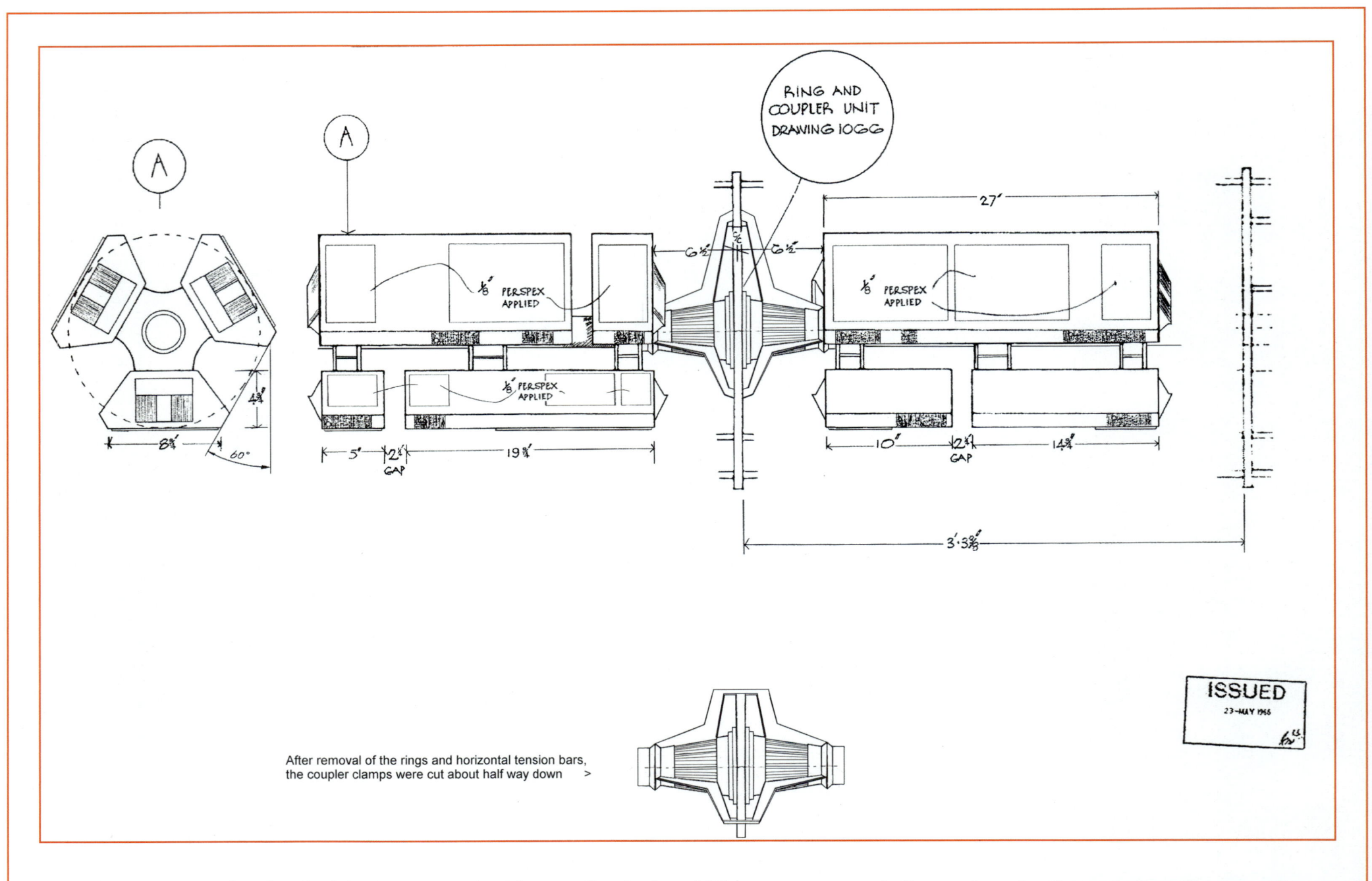

Mid-Section construction detail of the Storage Tanks. The coupling in the middle can open up and allow tanks to be disposed when they are no longer useable. This would be carried out when Discovery begins return from Jupiter mission to save fuel.

Close-up of the mid-section tanks and couplers with Communications antenna.

The Discovery's Gaseous fission plasma reactor

Thomas F. Widner, General Electric's manager of advanced nuclear programs was the primary consultant on the gas nuclear or 'Cavradyne' reactor. It confines gaseous fission fuel magnetically (MPD), and electrostatically in the reactor so that it would not touch (and melt) the reactor walls. The benefit of the gaseous reactor core is that it extracts electricity magnetohydrodynamically (MHD), or with simple direct electrostatic conversion of the charged particles.

The gas or vapor core is composed of Uranium Tetrafluoride with some ^{3}He added at the outlet to increase the electrical conductivity. The vapor core also has tiny UF_4 droplets in it. Since the space engine is not economical in the traditional sense, it also allows for a higher ratio of UF_4 to helium in order to increase the efficiency of direct conversion. The outlet temperature is raised to that of the 20,000 K range where the exhaust consists of fission-generated non-equilibrium plasma gas.

The open cycle propellant hydrogen is fed to the reactor, heated by the nuclear reaction in the reactor, and exits out the other end being MPD controlled, and by the helium in the stream. A side effect is that the propellant will be contaminated by fuel and fission products, and is mitigated by engineering the magnetohydrodynamics within the reactor. The uranium vapor is at most triple-ionized with a mass of 235 dalton (unit) Since the force imparted by the magnetic field is proportional to the charge on the particle, and the acceleration is proportional to the force divided by the mass of the particle, the magnets required to contain uranium gas are very large; this design has focused on fuel cycles that do not depend upon retaining the fuel in the reactor. Afterward, the uranium tetrafluoride is compressed by external means, thus initiating a nuclear chain reaction and a great amount of heat, which in turn causes an expansion of the uranium tetrafluoride. Since the UF_4 is contained within the vessel, it's only escape is through the outlet nozzles, resulting in a plasma wave moving through the container and out the nozzles.

The unique elliptical shaped front of the reactor coincides with the shape of the propagation waves generated by the reactor core. This in turn counteracts the waves, thus dampening it.

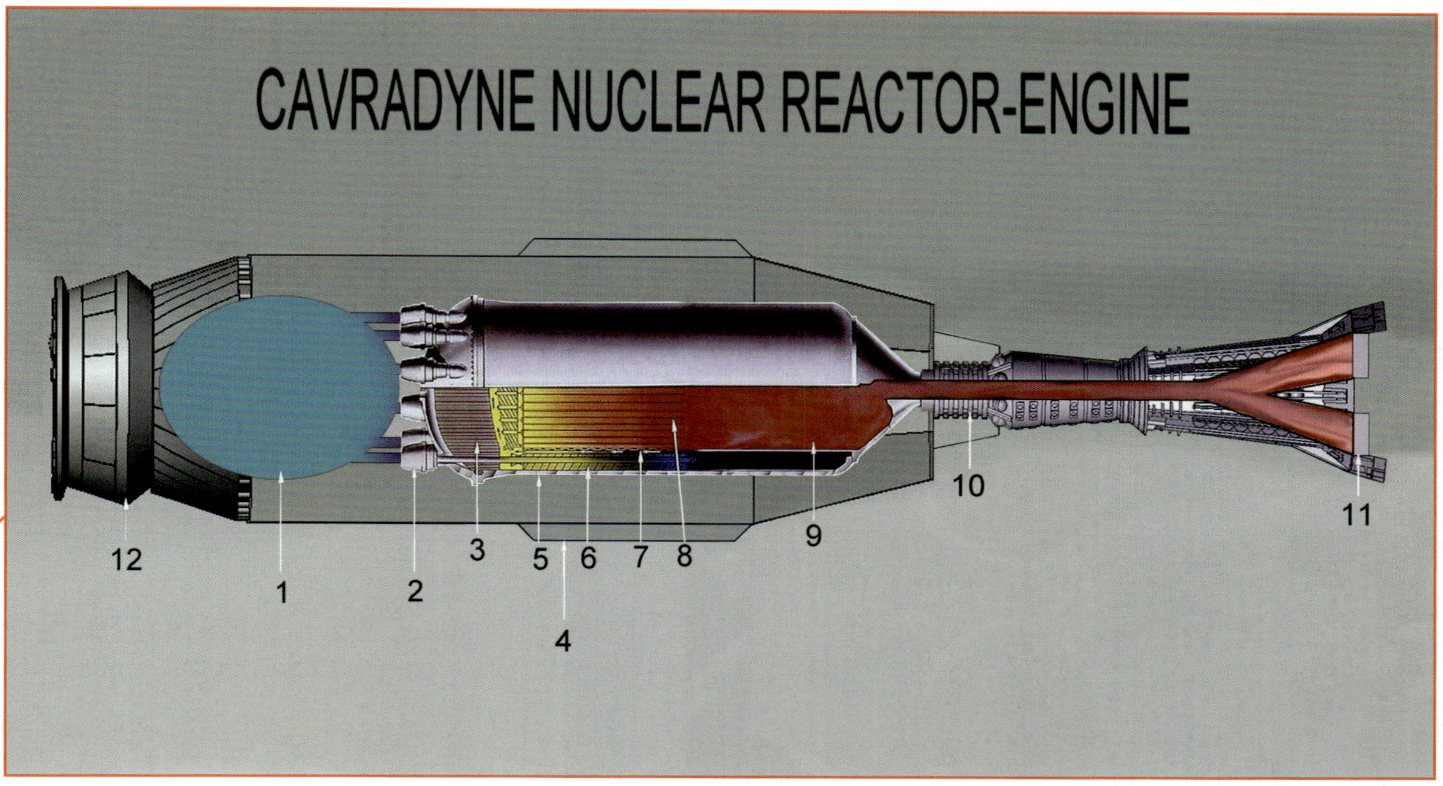

CAVRADYNE NUCLEAR REACTOR-ENGINE

1. Liquid Hydrogen tank
2. Turbine pumps
3. Graphite shield
4. 3 foot thick metal plate Magnetic Bottle Shields
5. Pressure vessel
6. Reflector
7. Radial support
8. Reactor
9. Gaseous cavity core of energized plasma.
10. Cooling fins
11. Thrust-vectoring vanes
12. Pressure plate – dampens vibrations.

Architectural craftsmen put final details on the main reactor core.

Three Cavradyne nuclear engines side by side. Two nozzles per engine are used to vector the exhaust.

Detail of the engine nozzle area.

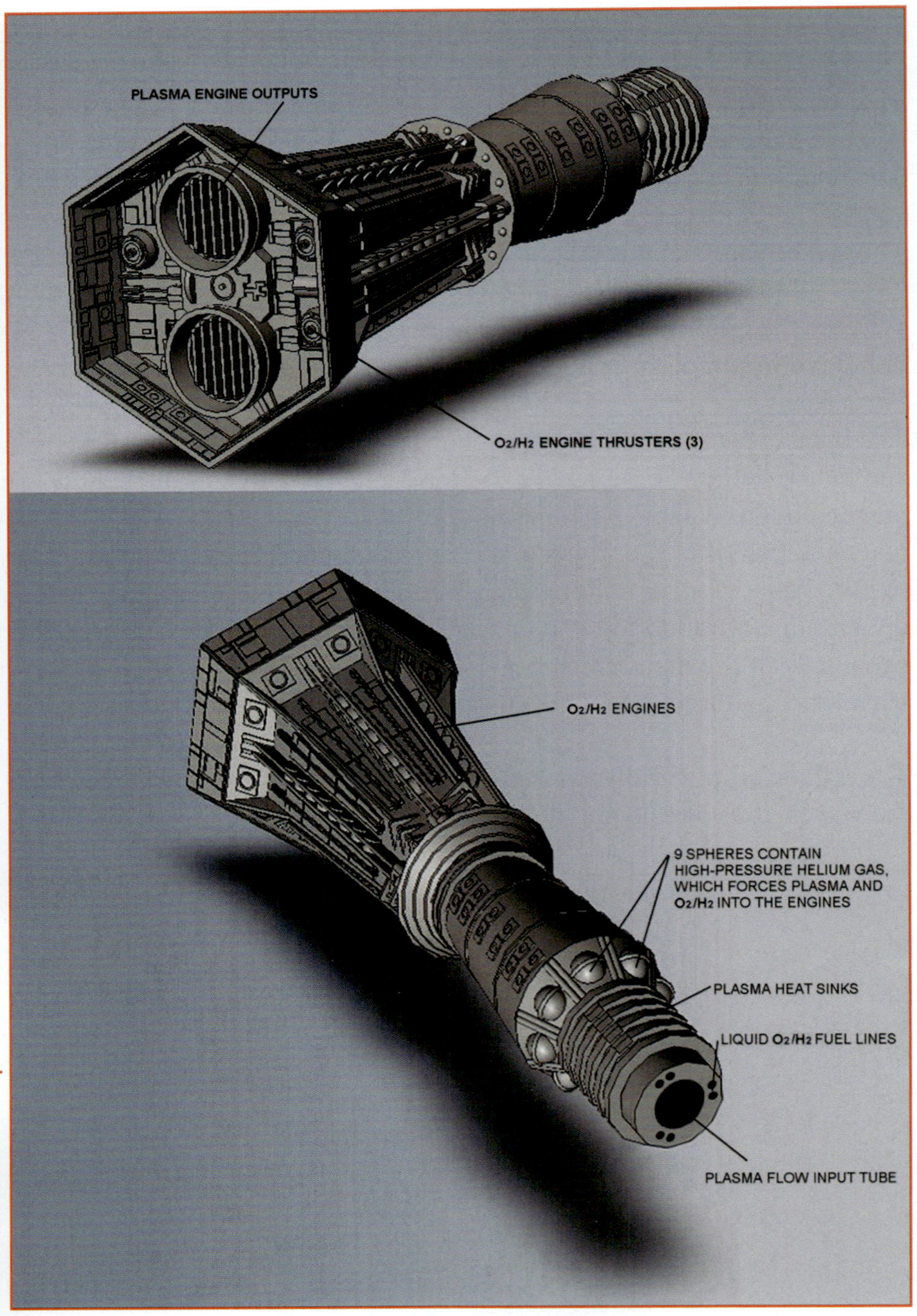

Detail of the thrust-vectoring area shows three O_2/H_2 chemical engine nozzles. These are used for controlled maneuvering in space close to planetary orbits.

Close-up of the engines at the rear of the Reactor. The rectangular blocks between each engine house spherical O_2 and H_2 tanks for the O_2/H_2 thrusters.

PLASMA ENGINE OUTPUTS

O_2/H_2 ENGINE THRUSTERS (3)

O_2/H_2 ENGINES

9 SPHERES CONTAIN HIGH-PRESSURE HELIUM GAS, WHICH FORCES PLASMA AND O_2/H_2 INTO THE ENGINES

PLASMA HEAT SINKS

LIQUID O_2/H_2 FUEL LINES

PLASMA FLOW INPUT TUBE

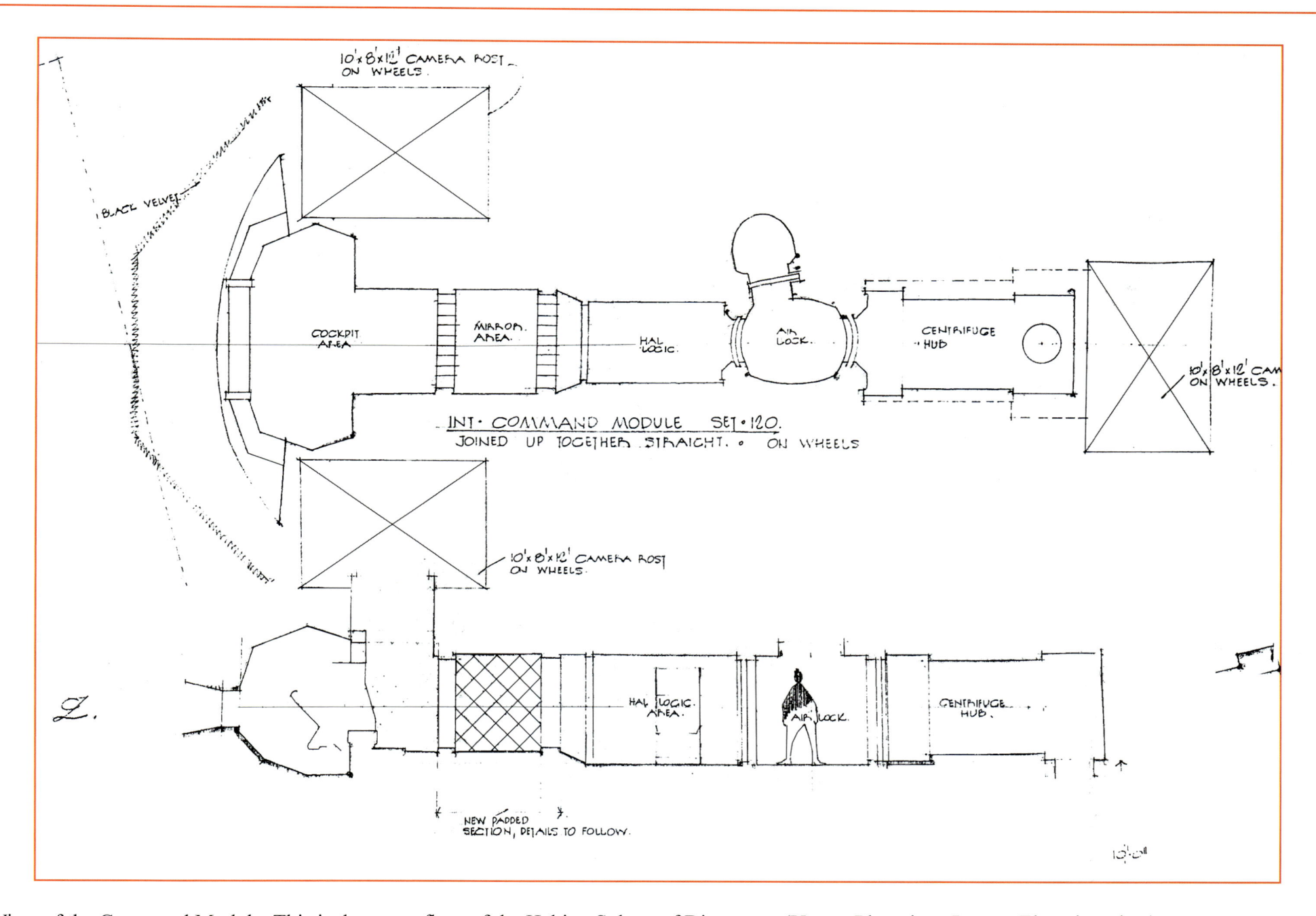

View of the Command Module. This is the upper floor of the Habitat Sphere of Discovery. (Upper, Plan view; Lower, Elevation view)

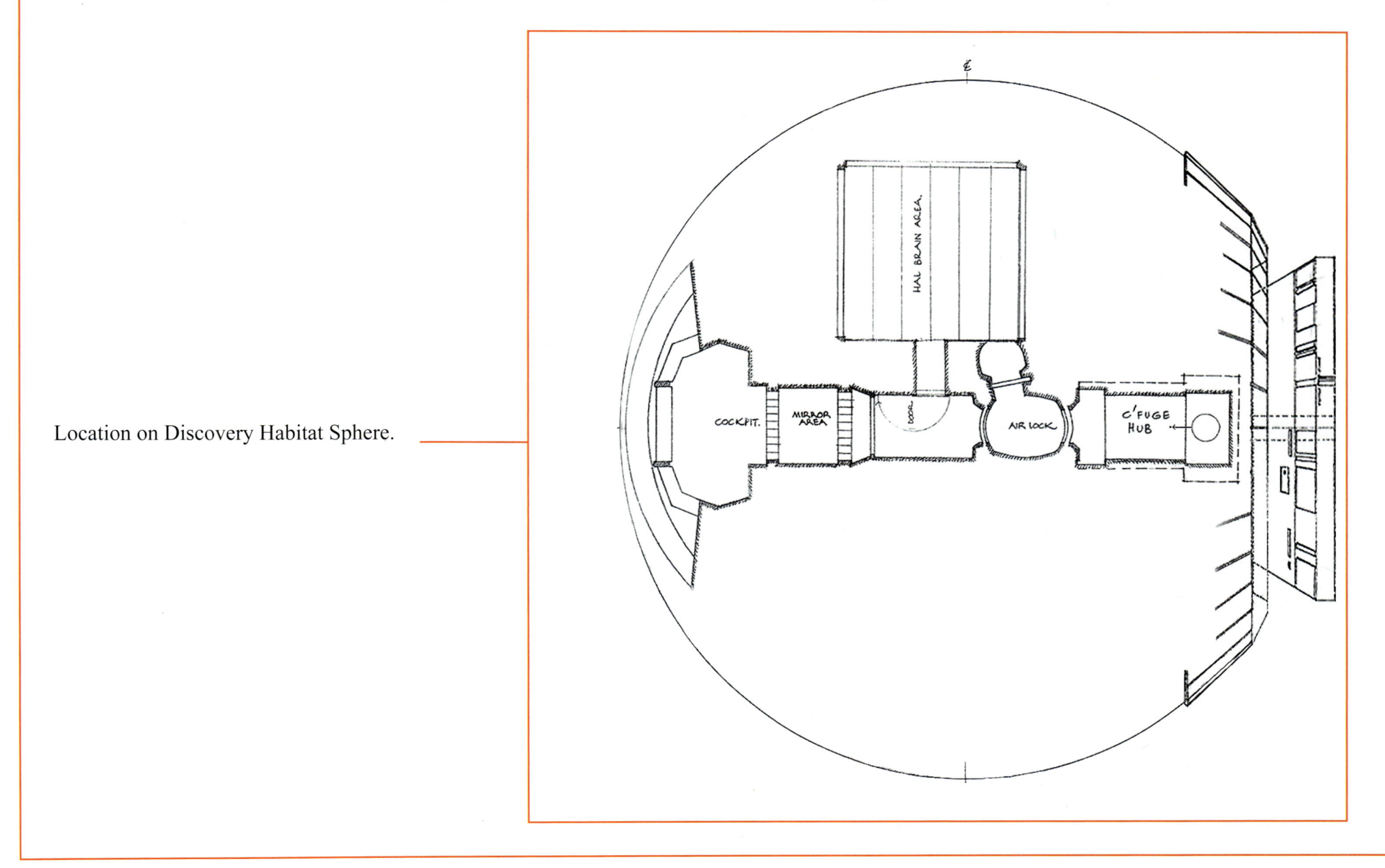

Location on Discovery Habitat Sphere.

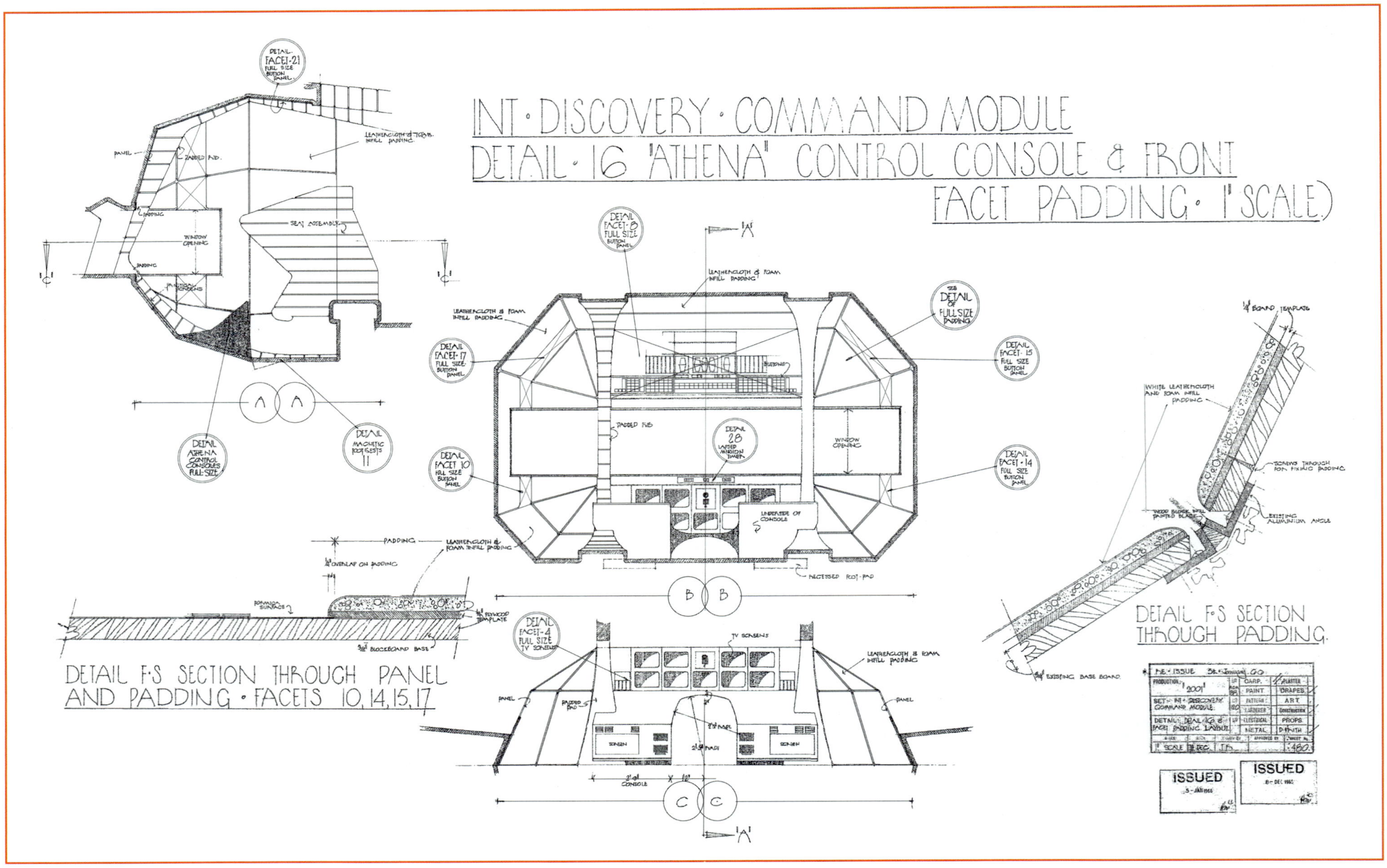
INT · DISCOVERY · COMMAND MODULE
DETAIL · 16 'ATHENA' CONTROL CONSOLE & FRONT
FACET PADDING · (1" SCALE)
DETAIL F·S SECTION THROUGH PANEL
AND PADDING · FACETS 10, 14, 15, 17
DETAIL F·S SECTION
THROUGH PADDING
ISSUED
ISSUED

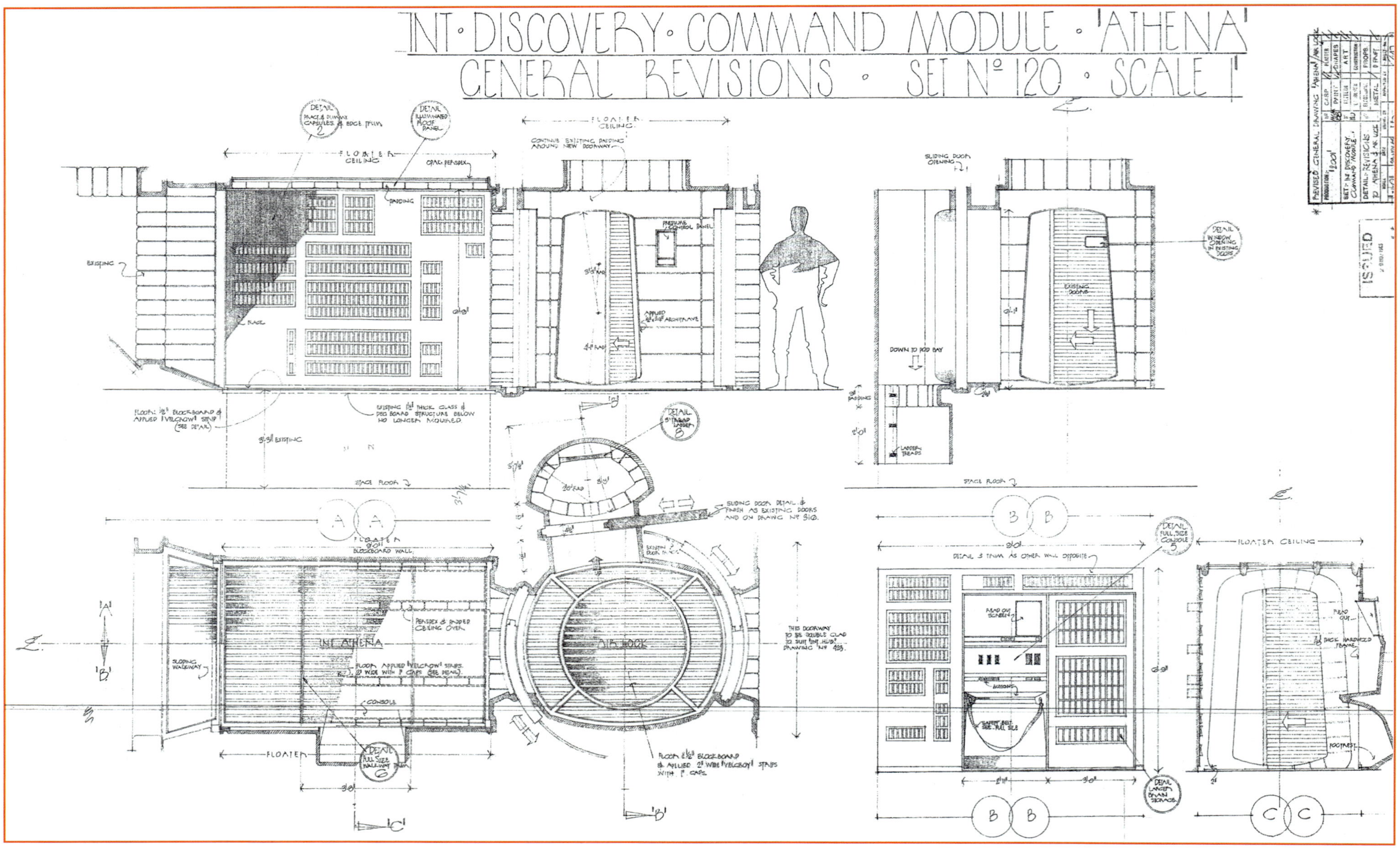
INT · DISCOVERY · COMMAND MODULE · 'ATHENA'
GENERAL REVISIONS · SET Nº 120 · SCALE 1"
ISSUED

View of the command deck as seen through front windows. Floor is oriented 90° to the front window.

Corridor directly behind front two seats. An astronaut is preparing to enter the HAL 9000 'brain room.' The walls have a vast array of relays and manual switches that are integrated into the HAL 9000 computer.

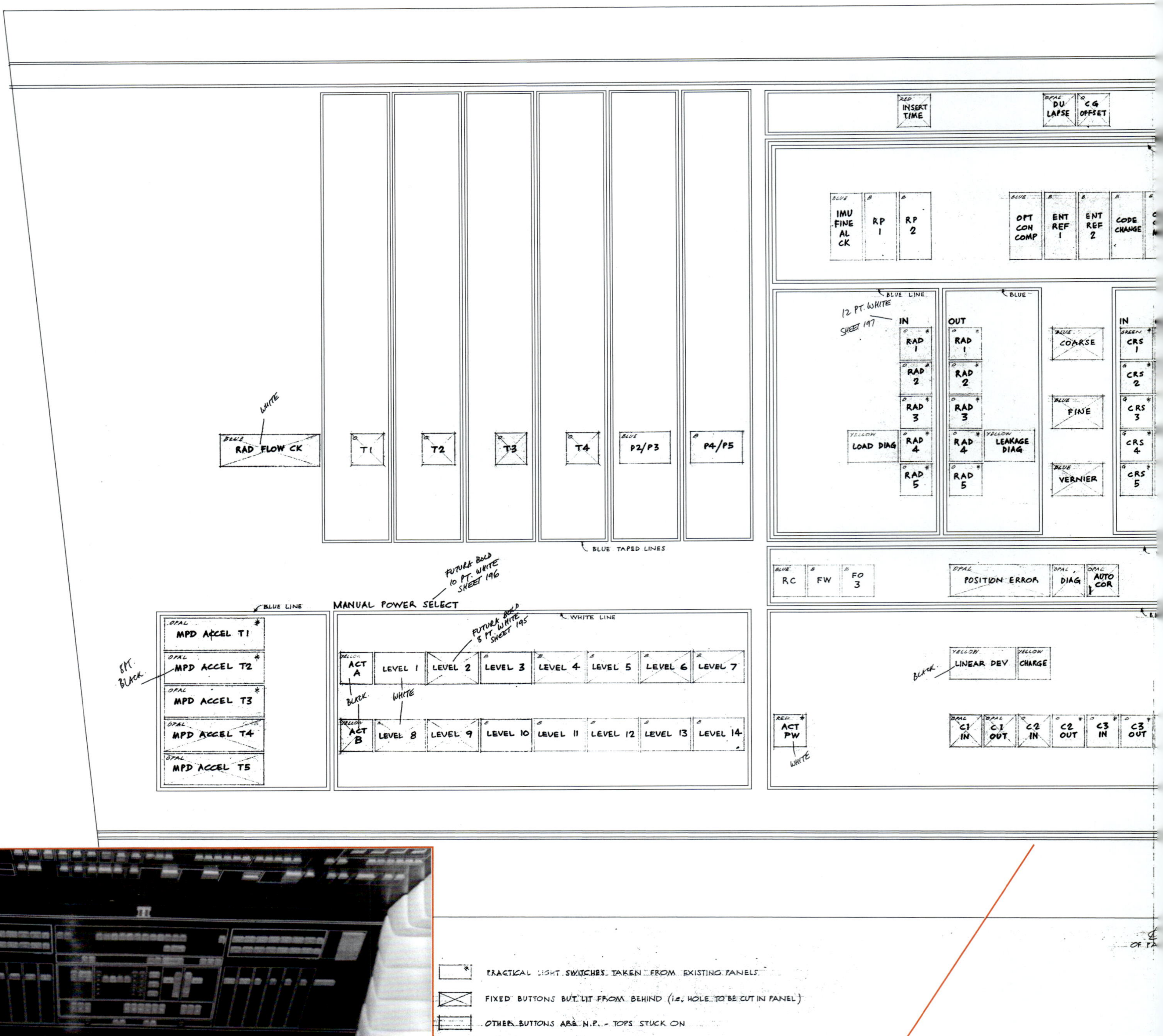

This panel, directly overhead of the two 'command chairs,' is a fine scale monitoring station of the Cavradyne reactor. It measures the state of the magnetic bottle (MPD) around the cavity core, the radiation levels, neutron levels, and user commanded tests of each subsystem. The HAL 9000 computer may alert the crew of a status change of the Cavradyne reactor and then manual overrides would occur here.

Ordway stands under the Reactor Control Panel.
Behind him is the front window of the Command Module.

1 PANEL ONLY REQUIRED

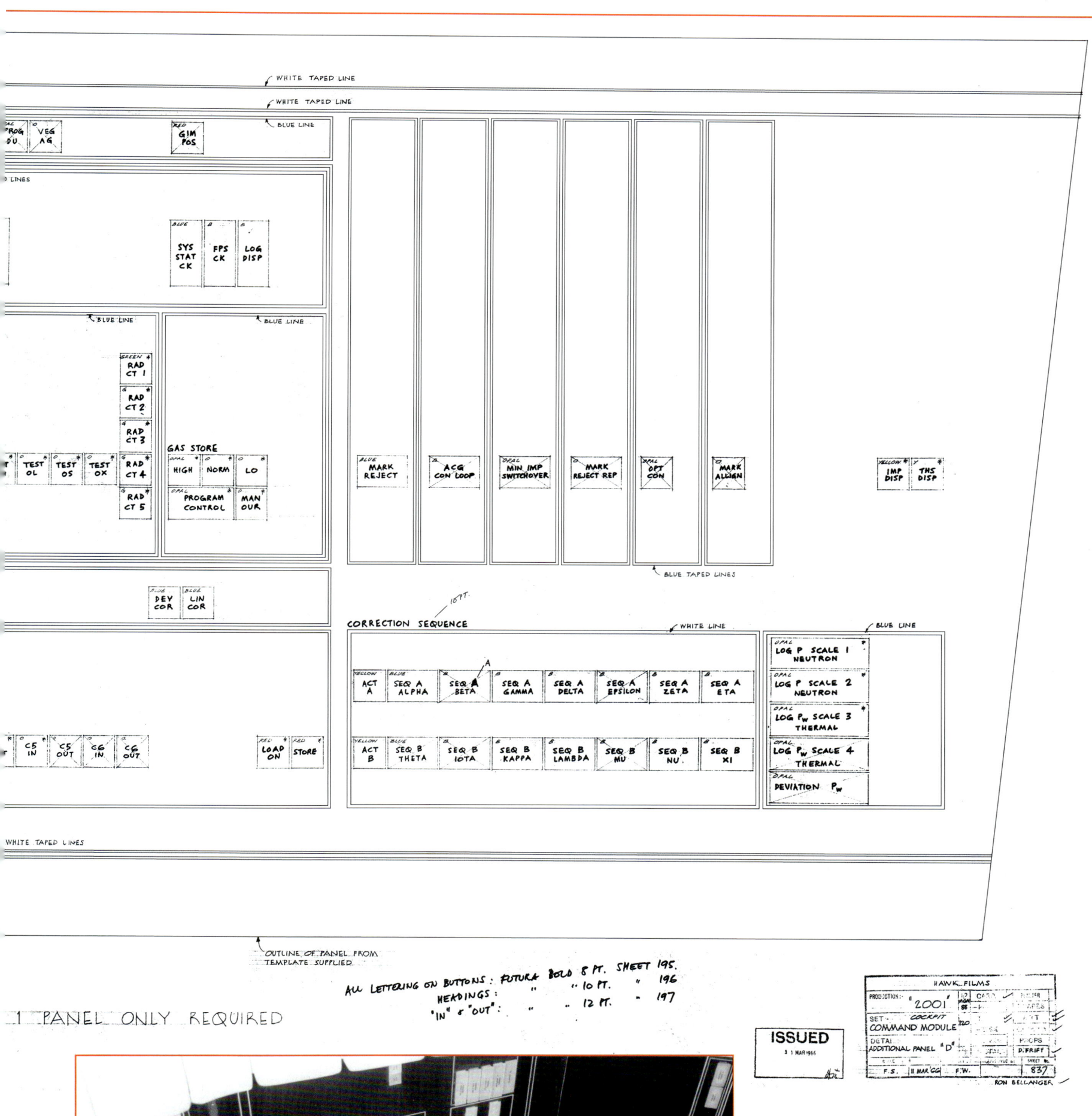

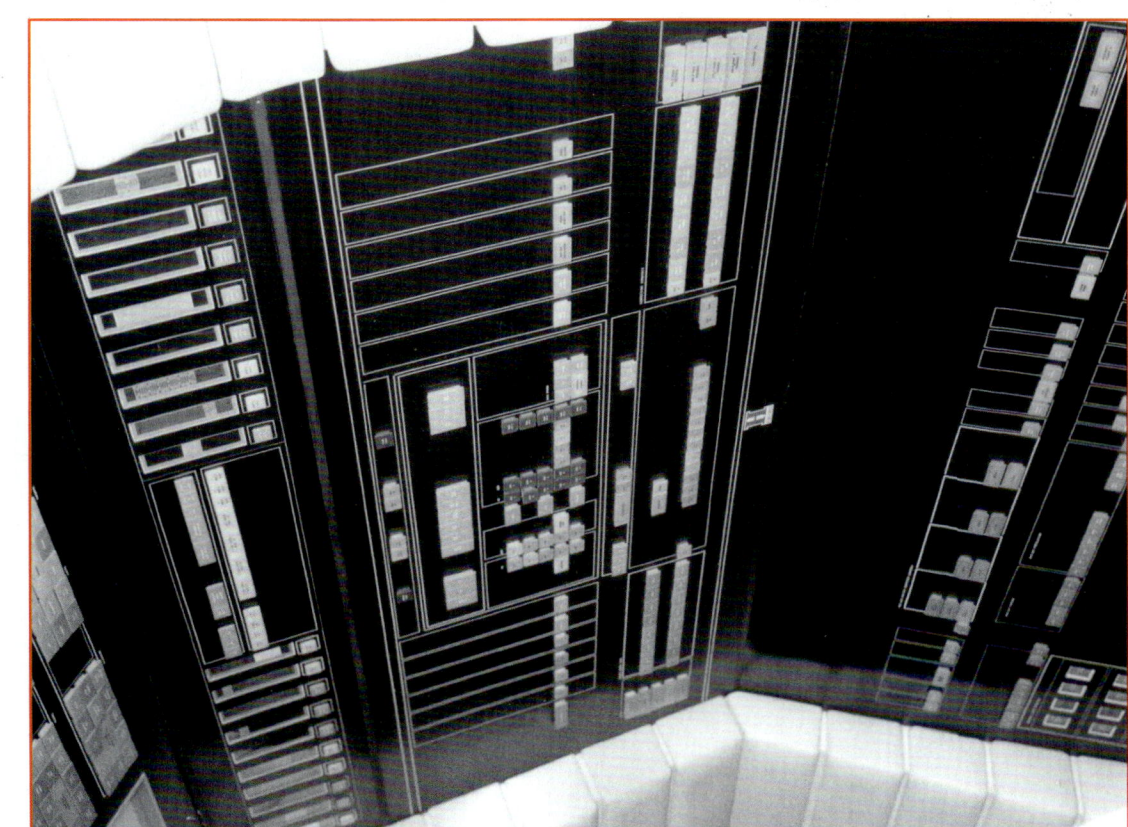

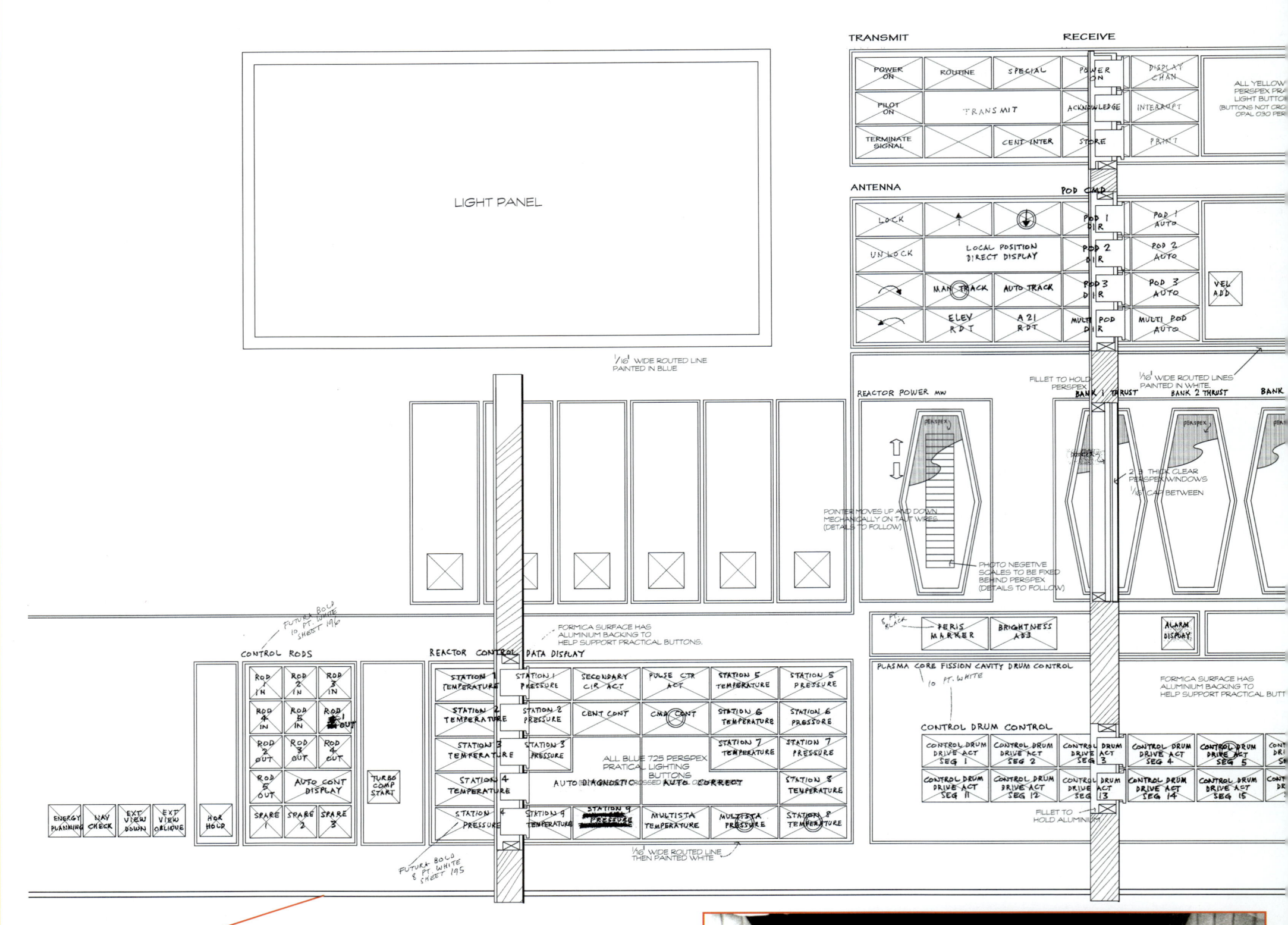

The Reactor Control Panel. This panel is the master controller for the Cavradyne engines. The center displays the thrust output of the 3 engines. From left to right, user inputs the reactor rods, monitors the various sensor points, then while fusion occurs, engages the cavity core drum one segment at a time. Reactor secondary controls monitor the containment of the reaction for the purpose of generating electricity to run the whole vehicle. ESPS buttons activate and display the electrical functions throughout. The top buttons (between light panels) are detailed controls of the AE-35 antenna mounted mid-ship.

Command Module looking through the front window.

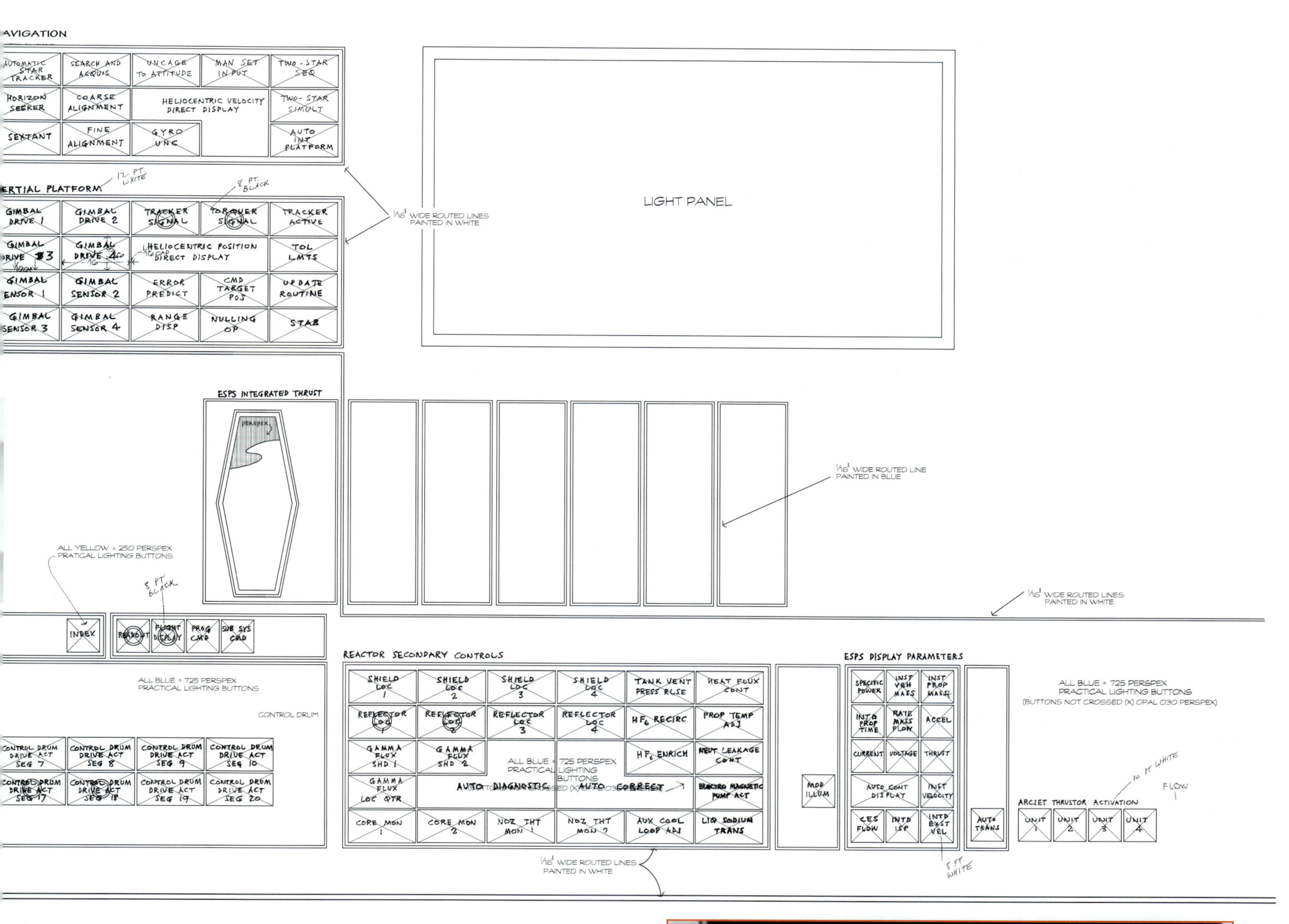

Close-up of the Command Chair has vernier engine controls on the left and right sides of the seat. The left controls an engine group. The right controls the maneuvering manually. For example: pitch up, roll left, roll right, etc. The button on the panel directly in front of the seat call up any part of the mission that has been digitally stored in the computer library. This is viewed on any of the eight monitors centered between the seats. Verbal commands to the HAL 9000 can also be used.

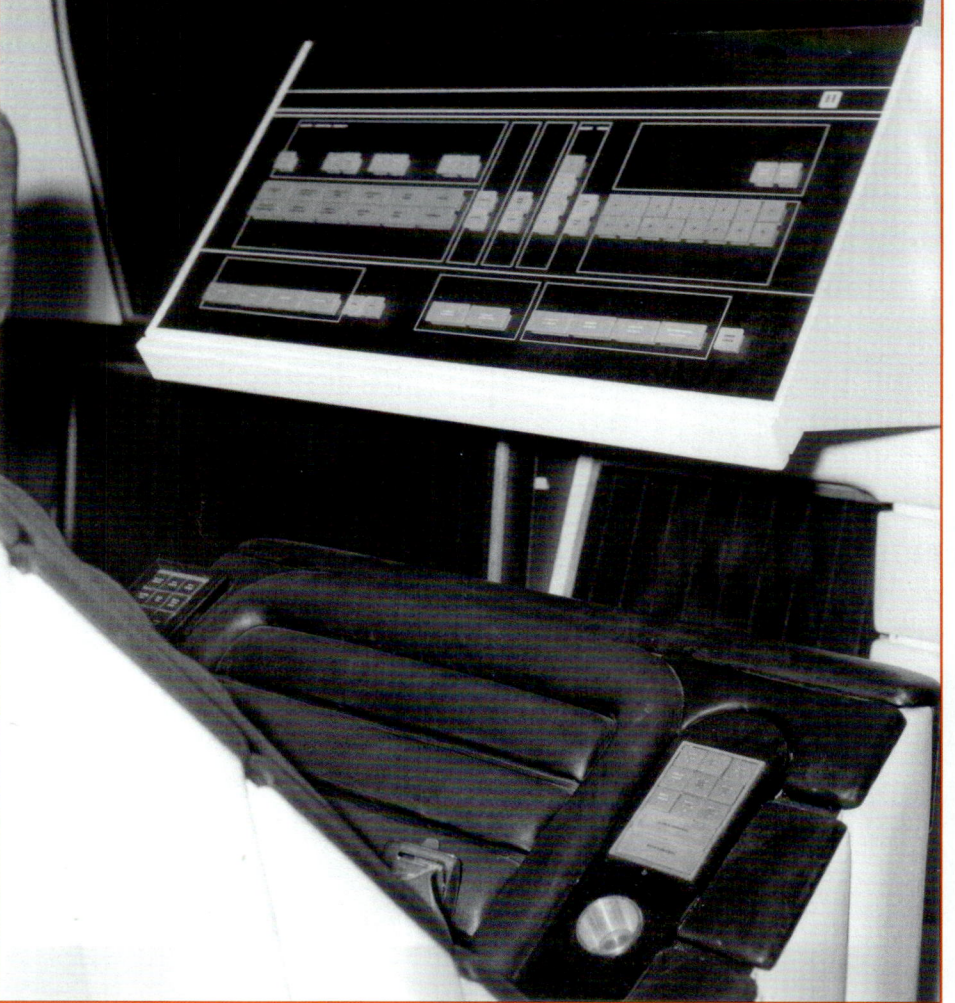

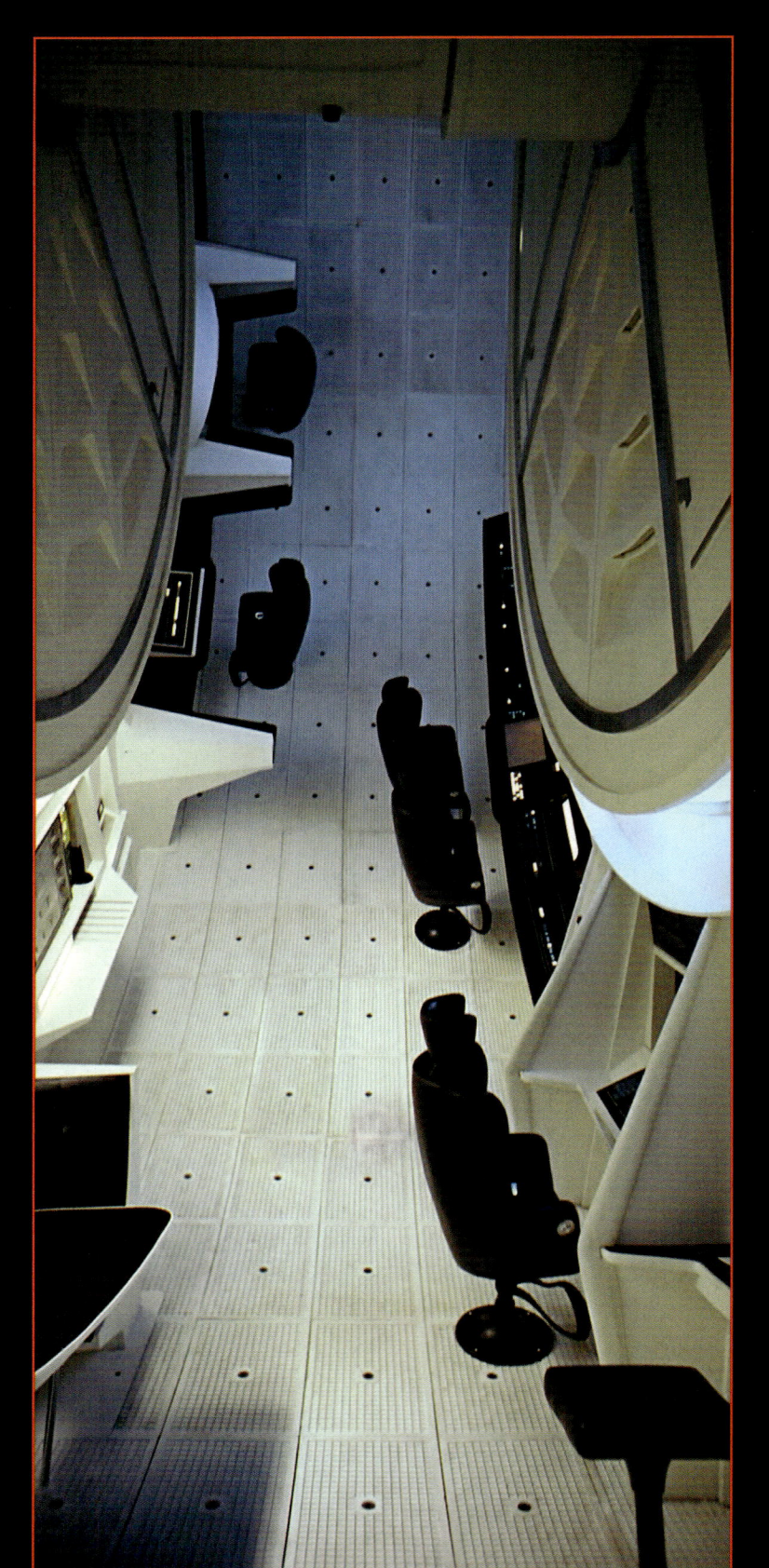

The Centrifuge

The centrifuge consists of rim-installed consoles, panels, screens, and devices. There is an automated kitchen; a ship-to-Earth communications center; a complete medical sections where the astronauts undergo regular automated checkups (results are displayed and recorded, diagnosis of deficiencies is given directly on a readout screen, and medicament or other treatment prescribed); an observatory, and a geophysical exploration module. The latter permits a wide variety of surface and subsurface experimentation to take place on an alien body, such as an asteroid or a moon. Since subsurface structure could be extremely important in the spaceship's investigatory program in the Jovian system, a drill is incorporated into a remotely controlled surface lander. Controls on the console include a depth selector, drilling rate selector, equipment calibration, recording and error analysis controls, and various screen and gauge indications of subsurface characteristics – formation type, formation content, well horizontal cross section, caliper (symmetrical curve representing a vertical cross section of the hole being drilled), sonde 'up' and sonde 'down,' recording, etc.

We had the option of putting the Centrifuge on for, say, one to two hours a day to produce up to 1.5 g, or permanently have it rotate to provide about 0.2 to 0.3 g. We chose the latter. There was, of course, the problem of Coriolis forces, which on small diameter wheels would cause dizziness to astronauts walking along the rim. Calculations showed that a centrifuge should be at least 300 ft in diameter to reduce to acceptable levels the inconveniences caused by the Coriolis forces, but such a diameter was beyond the capabilities of the M-G-M British Studios – and our budget. So we never really mentioned the diameter of the wheel with which we had to work; in fact, there was no purpose to reveal the measurements at any time. Visual appearances were what counted.

Meticulous attention was paid to all the modular elements making up the Centrifuge, and thousands of hours of research and development went into them to ensure utmost accuracy and realism. The modules were:

Side 1	Side 2
Hibernation*	Hibernation
Health monitoring	Medical checkout
Hibernation	Hibernation
Gymnasium	Gymnasium
Storage	Shower and hand basin
Entertainment	Dining area
Communications	Kitchen
Storage	Nuclear reactor controls
Flight instrumentation	Telescope
HAL multiscreen area	Geophysical console
Lockers	Lockers
Pressure room**	Pressure room**
Toilet and hand basin	
Ladder	

* All hibernacula were also designed for sleeping by non-hibernating astronauts.

** Emergency pressurized area and 'storm' shelter for protection against solar storms.

Engineering and construction was provided by Vickers-Armstrong Ltd.

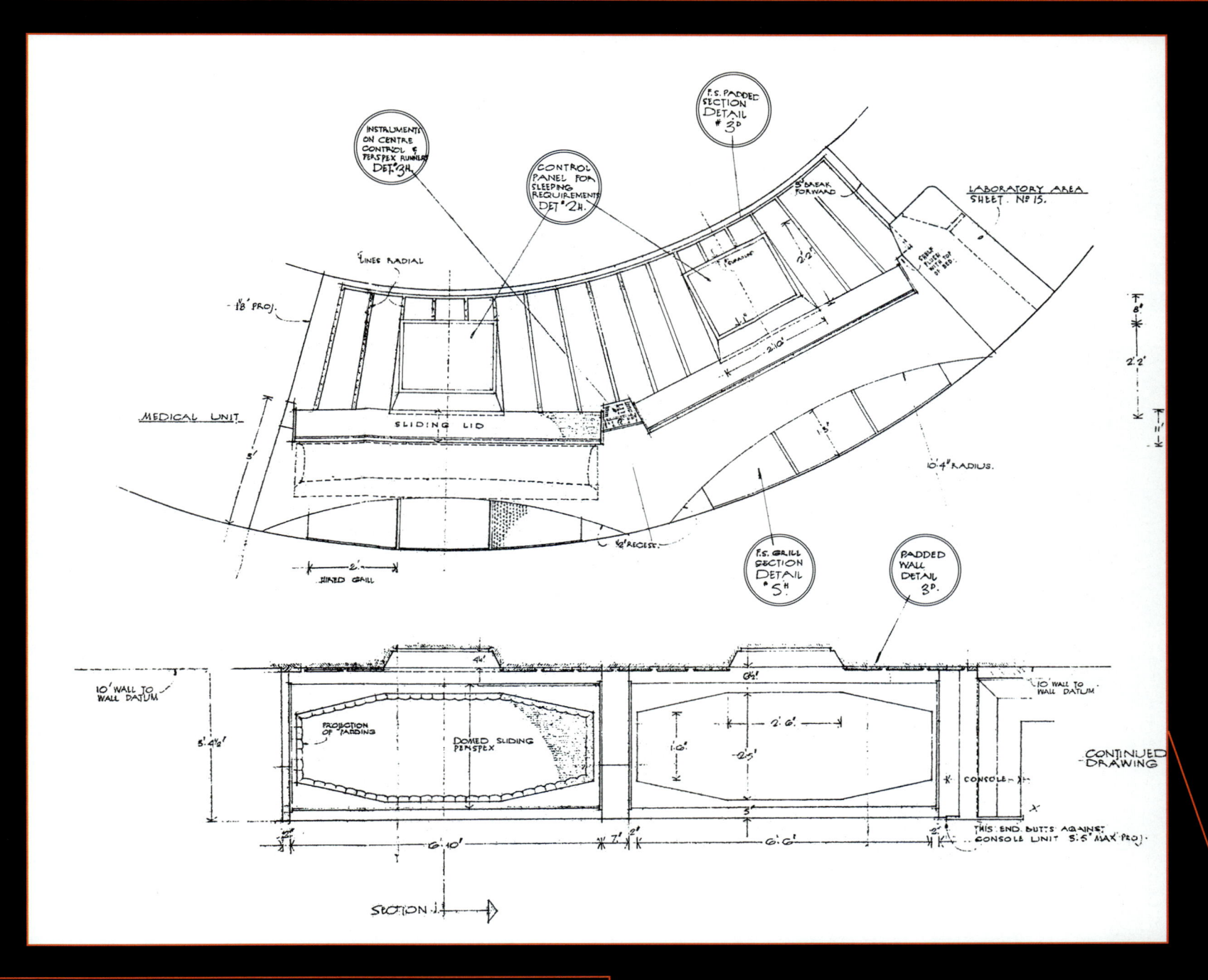

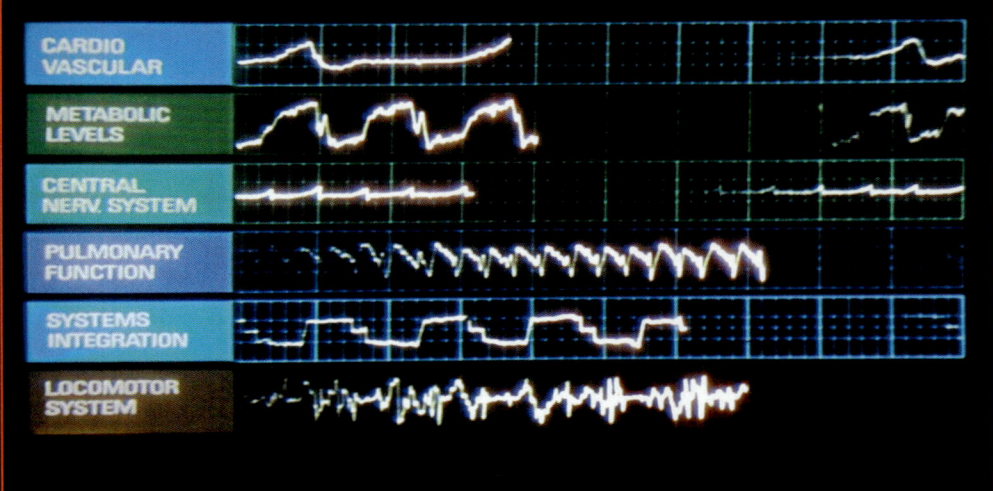

Two of the five 'Hibernacula'. Three are used for Hibernation, and the other two for the 'wake' astronauts for daily sleep routines.

Techniques for placing the three scientist-astronauts in hibernation inside the centrifuge were worked out with Dr. Ormand G. Mitchell of the New York College of Medicine, DR. A.T.K. Cockett of the Harbor General Hospital in Los Angeles, and Drs. K.G. Williams and Peter Scott of the Medical Division of Vickers Limited in London. Each astronaut is monitored during the long Earth-orbit to Jupiter-orbit flight with respect to locomotor system; central nervous system; cardiovascular activity, e.g., heart rate, myocardial state, cardiac output, blood pressure, capillary exchange, lymph flow; systems integration, e.g., endocrin control, neuro control and balance; metabolic levels, e.g., acid/base balance, renal functions, nutrient input, deep body temperature, hepatic function; sensory activity; pulmonary function; and such direct hibernation controls as hypothalamus stimulation, temperature rise A (heart rate), temperature rise B (respiratory rate), sugar enrichment, thyroxin control, and vibrator.

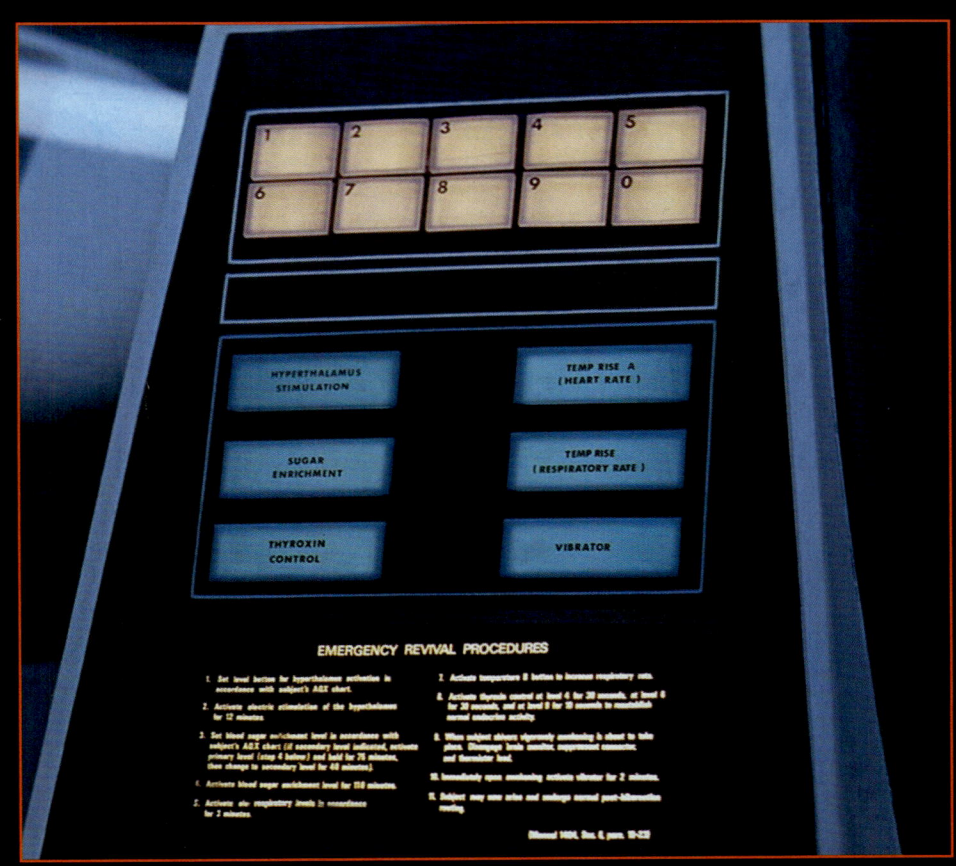

The sleeping crew members are programmed to emerge from the hibernating state at a specific time in accordance with a specific routine. Nevertheless, in case of emergency or unforeseen circumstances, they can be brought out through special instructions by the computer or by manual override.

A digital synthesiser is available for entertaining the crew during the long mission. A library of thousands of songs and sheet music can be called up on the display.

The Communications module is linked to the large AE-35 antenna and two twin aperture antennae.

Entertainment

Communications

Storage Entertainment
Gymnasium Communications
  Storage
  Instrumentation
Hibernation
  HAL Screens
Health Monitors
  Lockers
  Pressure Room
Hibernation
  Toilet & Hand Basin
Health Monitors Hibernation Instrumentation
  Ladder

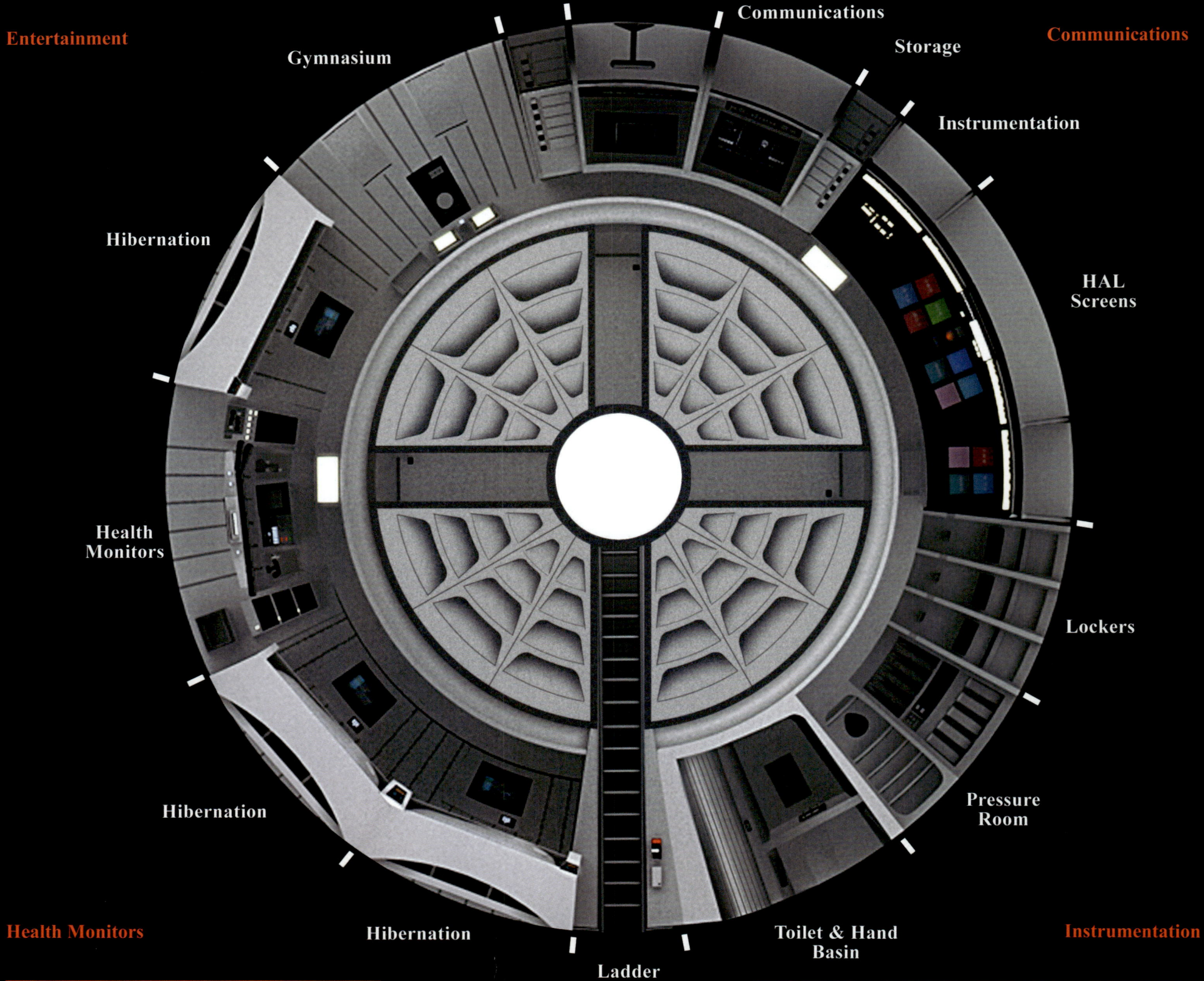

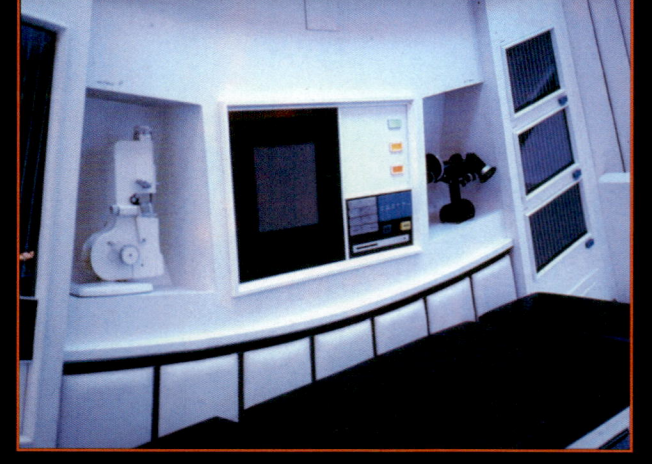

Instrumentation. This module controls ventilation in the Centrifuge, oxygen levels, and purging of CO_2 out of the area.

The Health Monitor Module is for routine checkups of the 'wake' crew. Three panels to the right contain medicines and basic emergency and first-aid supplies. Microscopes for medical evaluation designed by Bausch & Lomb. Directly above is a sun-lamp for crew relaxation and tanning, which helps Vitamin D production from the absence of sunlight.

HAL Multiscreen left panels

HAL center panel

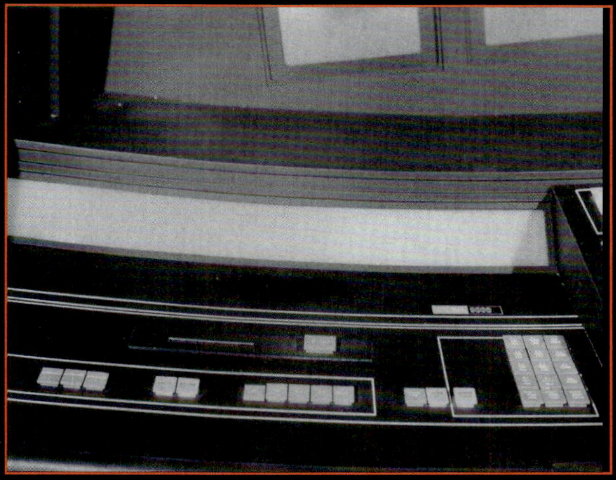

HAL 9000 multi-screens, middle and middle-right. Crew spends most of its time communicating with HAL here. Any camera view inside and outside the Discovery can be manually or verbally seen here. Intercoms, data and complete ship systems monitoring is generally monitored here. The panel directly in front of the 'AV' interface is used for leisure games; for example, chess, checkers, solitaire, etc. Button panels designed by RCA.

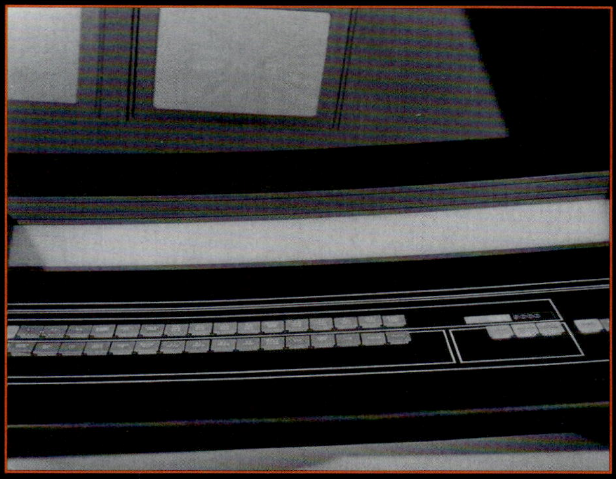

HAL Multiscreen middle left with printer slot

HAL right panel

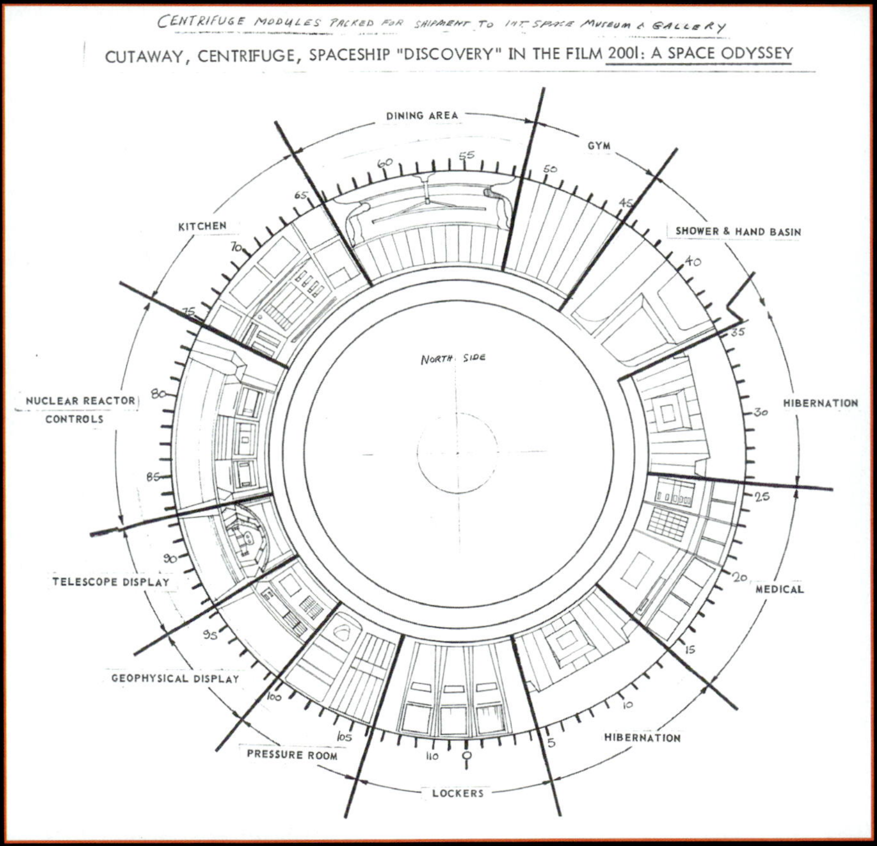

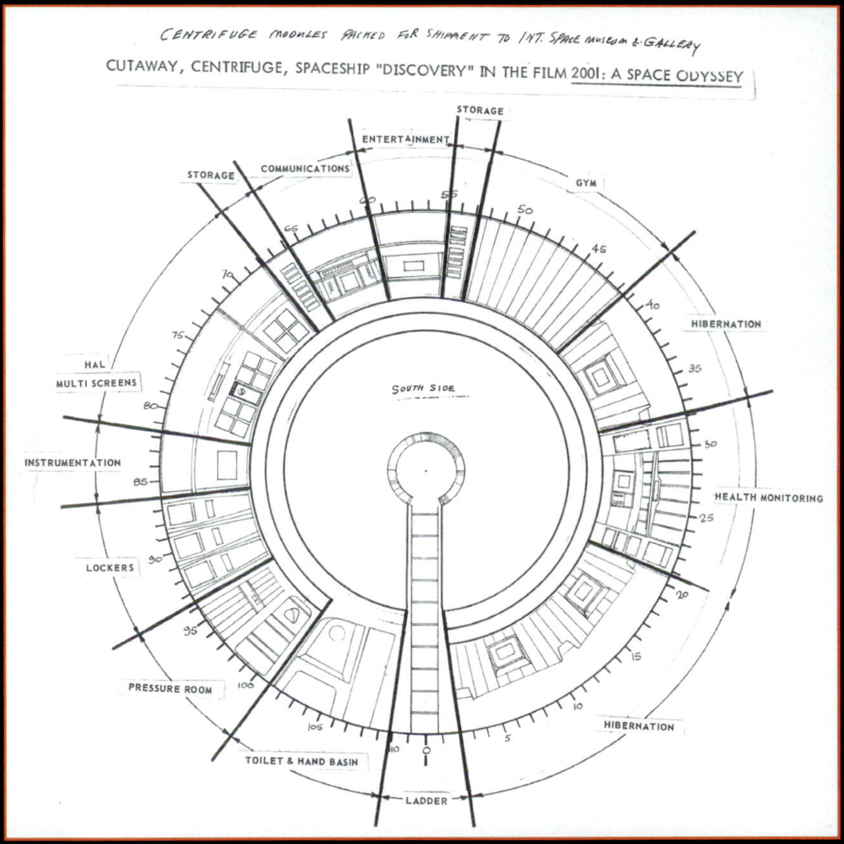

Original schematics of the centrifuge. The South Side drawing (right) is inverted as though it was originally sketched by someone standing outside the set. The actual set layout is depicted accurately in the color diagram on the opposite page.

After long working sessions, we completed our design of the astronomical or telescope module, with its telescope viewing piece, controls and displays. All manner of data could be read out upon command, including azimuth rate increment, elevation rate increment, declination, right ascension, differential coordinates, sidereal time, slewing rate, beam width, spectral dispersion range, and the like. Controls were incorporated to permit selection of any of four scopes, to change focus, to change magnifications, to make adjustments in right ascension and declination, to change alignments, to modify drive rates, etc. Direct screen readout was possible with magnifications of 20, 50, 250, 500, 1000, and 3000. Besides optical telescopes, Discovery had cosmic ray telescopes and other devices aboard to measure such phenomena as solar and galactic cosmic rays, the hydrogen corona formed around the planet by back scattering of Lyman-alpha radiation, and the particles trapped within Jupiter's magnetic field.

Telescope ports on outside of habitat sphere

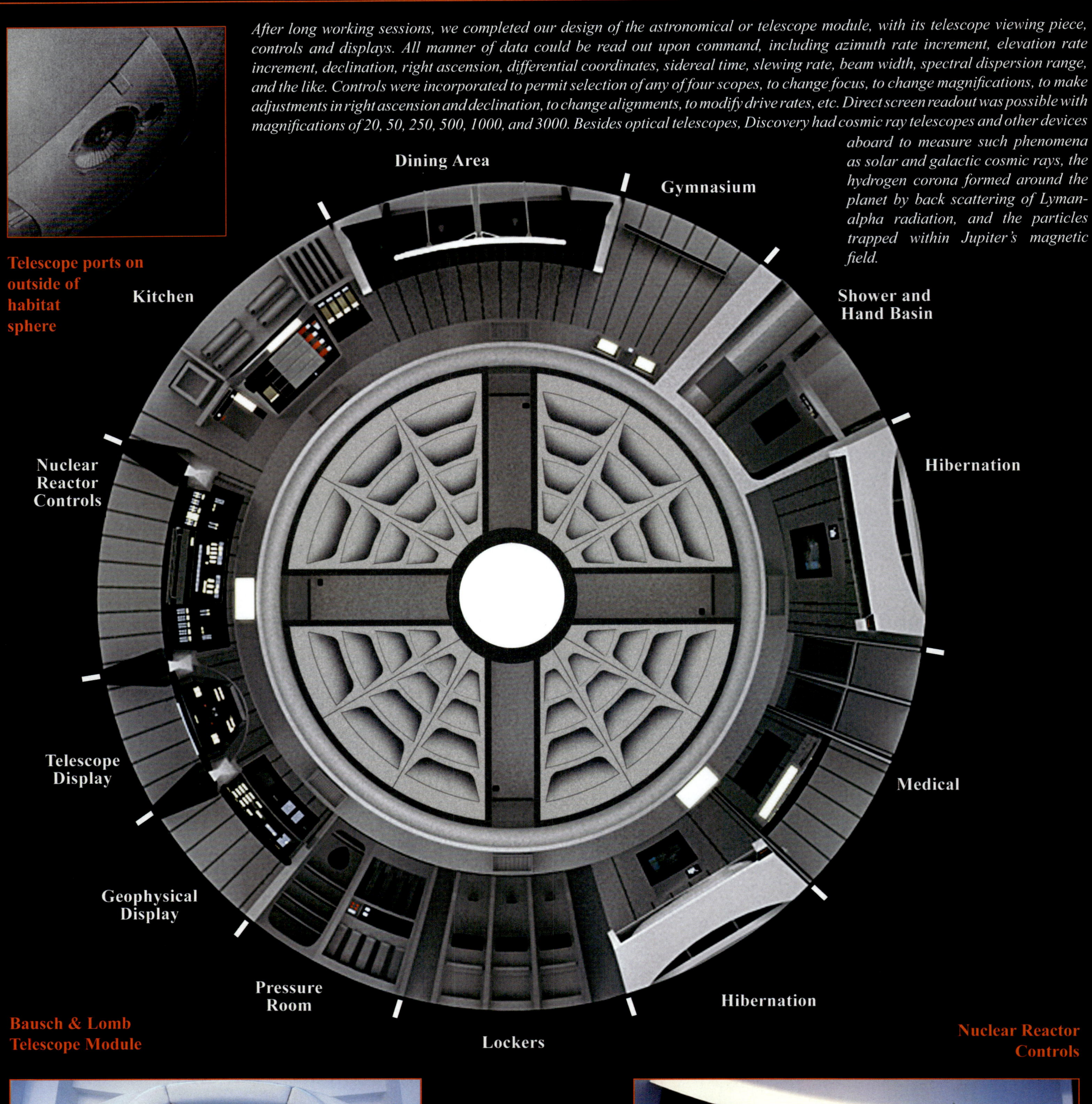

Dining Area

Gymnasium

Kitchen

Shower and Hand Basin

Nuclear Reactor Controls

Hibernation

Telescope Display

Medical

Geophysical Display

Pressure Room

Hibernation

Lockers

Bausch & Lomb Telescope Module

Nuclear Reactor Controls

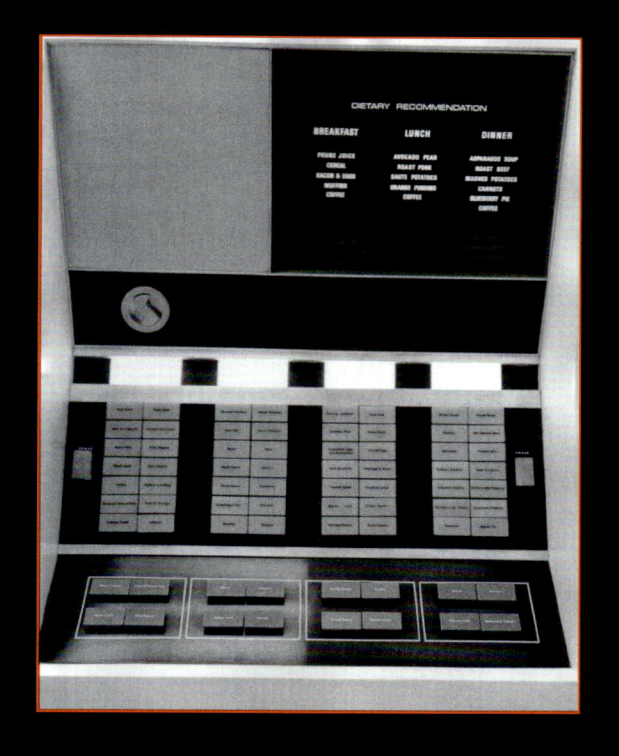

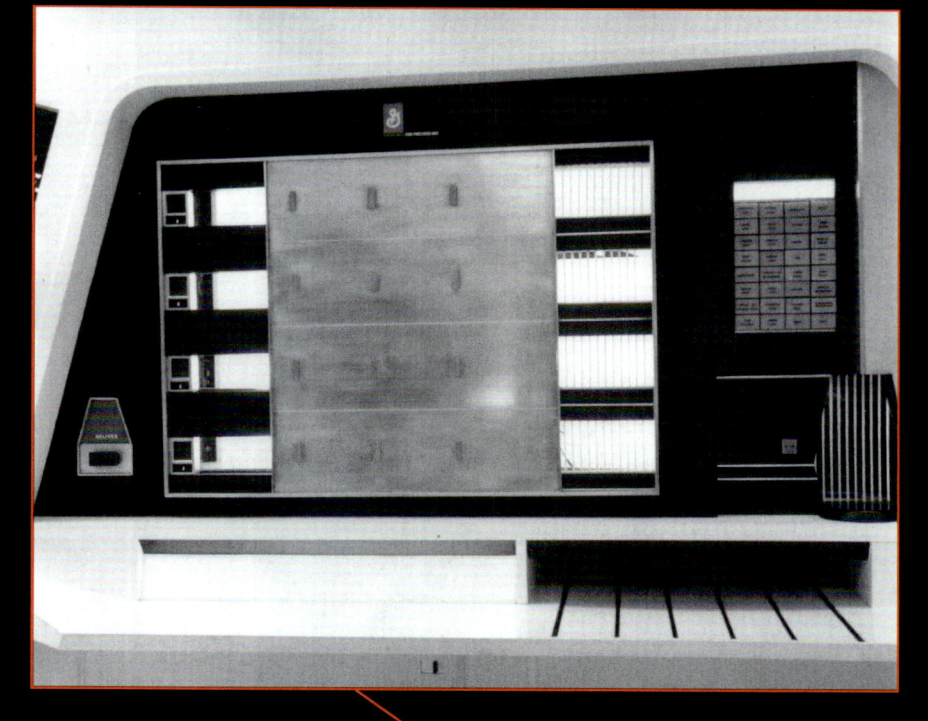

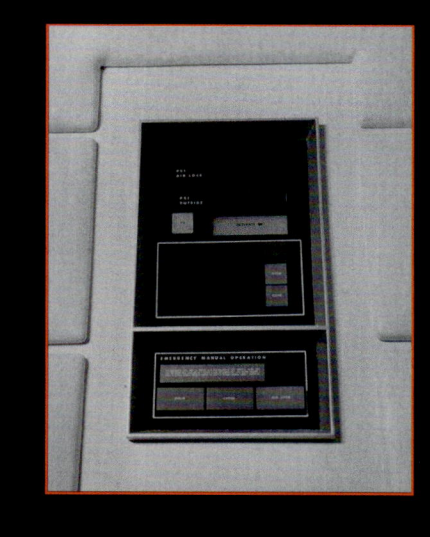

The Pressure Room is entered via this control pad. This area protects the crew from radiation or meteor bombardment.

In addition to routine vehicle and computer output data, special information can be transmitted from the *Discovery*, particularly that resulting from scientific experiments en route. These might include micrometeorite density as a function of distance from Earth (especially in the asteroid belt); location and study of asteroids heretofore undiscovered; and, at the Jupiter target, probing of the planet and its moons. There is a complete on-board astronomical observatory, designed with the cooperation of the Royal Greenwich Observatory, England, as well as instruments aboard to determine the population density and distribution of bodies from dust size upwards, and albedo and thermal flux detectors, and the like. If an individual asteroid or moon is to be probed, or a comet investigated by a small lunar lander, the Schlumberger designed geophysical console is called into use. Thus, it might be desirable, in a noninterference scientific investigation, to place a probe on a small asteroid to determine the nature of the subsurface and work out the microscopic structure. The Schlumberger equipment aboard *Discovery* permits a wide variety of surface and subsurface experimentation to take place. Since subsurface structure could be extremely important in the spaceship's investigatory program, a drill is incorporated into a surface lander. Controls on the console include a depth selector, drilling rate selector, equipment calibration, recording and error analysis controls, and various screen and gauge indications of subsurface characteristics, formation type, formation content, well horizontal cross section, caliper (symmetrical curve representing a vertical cross section of a hole being drilled), sonde "up", and sonde "down", "recording" etc. Diagnostic information is initially displayed as it is fed from the computer.

The Geophysical Module is used for inputting data from surface maps of planets and moons; which includes magneto surveys, seismology, surface reflection and absorption for mineral and energy exploration. Data would come from the EVA pod analyses and the telescopic data. Digital level meters on the horizontal surface measure trace amounts of radiation. Digital level meters (right side of horizontal surface) measure the independent levels of trace elements within a rock sample.

The Kitchen Module, designed and engineered by General Mills, supplies the crew with meals from a choice of menus. Food which is dry stored, is rehydrated and heated by infrared flash cooking. Plastic plates, cups and non-magnetic metal cutlery is disposed through a recycle chute on the right side. Here it is washed and sanitized automatically and then re-fed through the unit.

A shower and hand basin are available for the crew in the main centrifuge.

The Medical console displays and monitors a detailed view of medical data of the 'Hibernauts', the sleeping crew members.

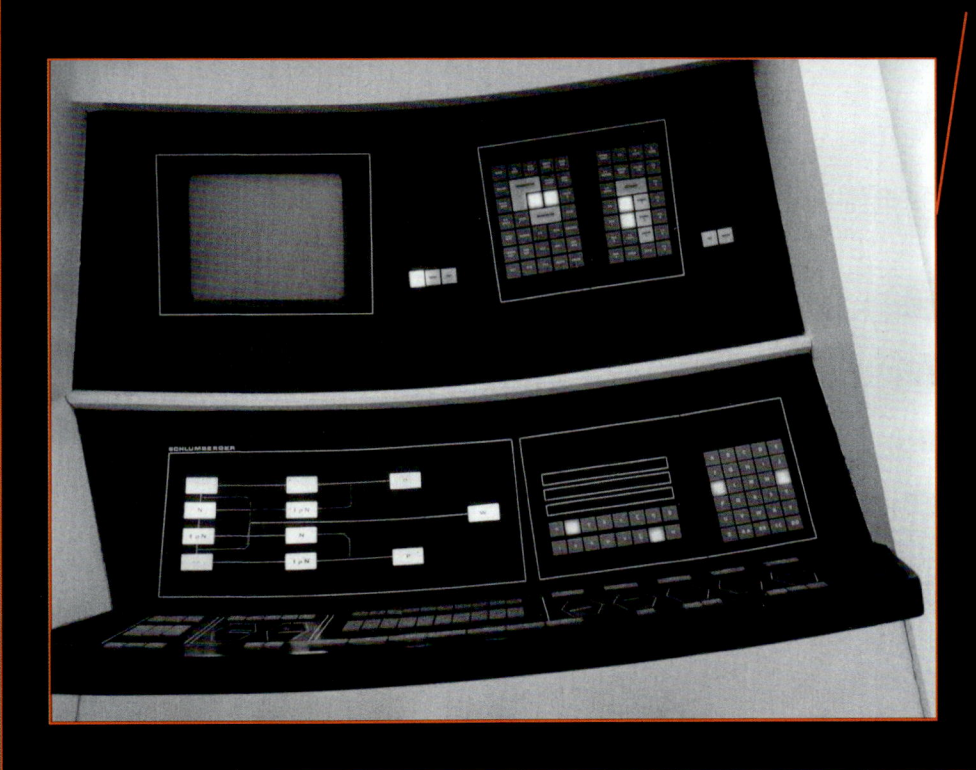

61

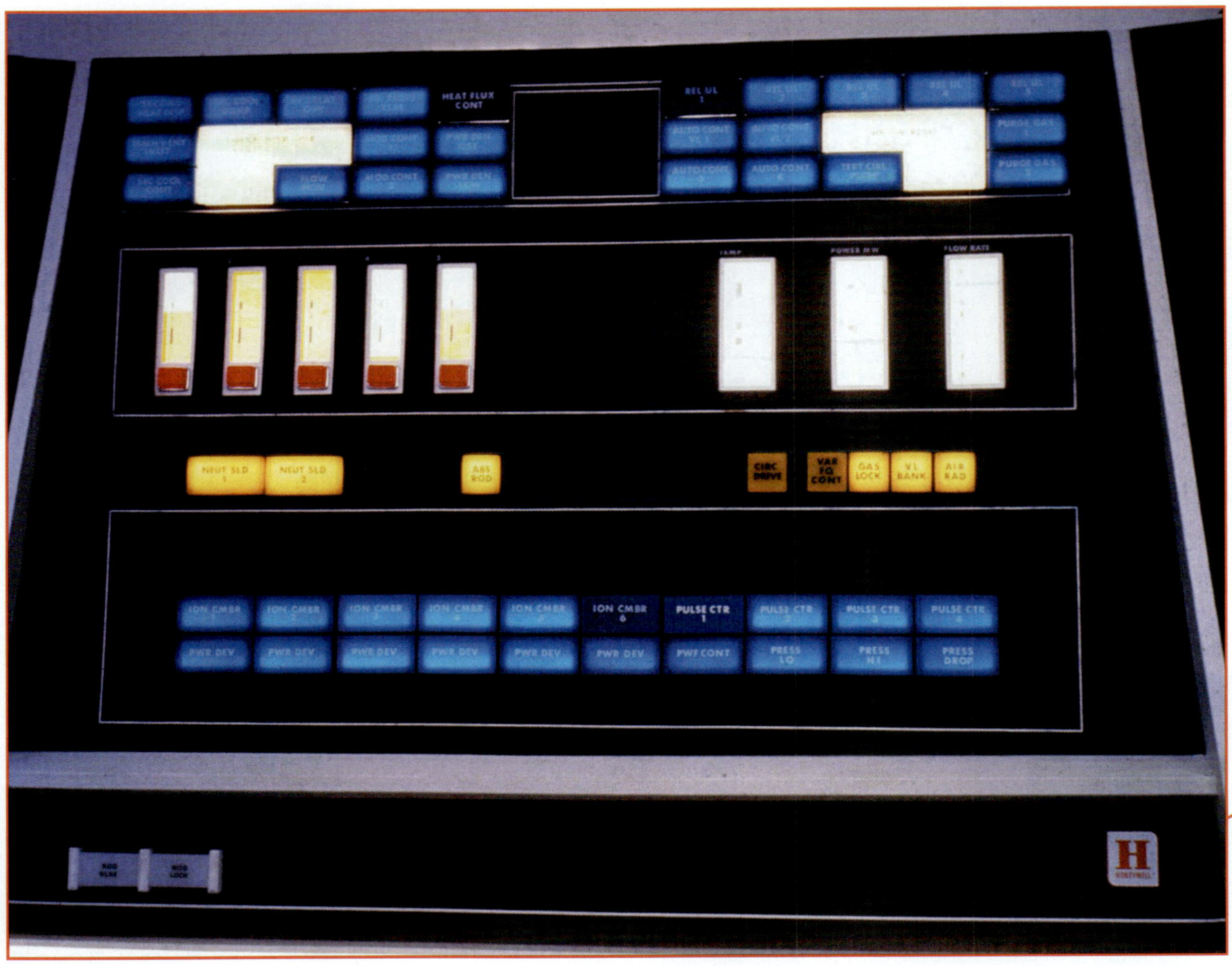

The central Reactor Control Panel is a back-up to the main panel located in the Command Deck. This one is used primarily after initial launch from orbit, considering the crew spends almost all their time in the Centrifuge during flight time.

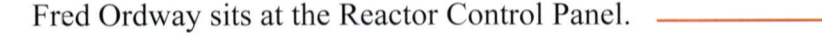

Fred Ordway sits at the Reactor Control Panel.

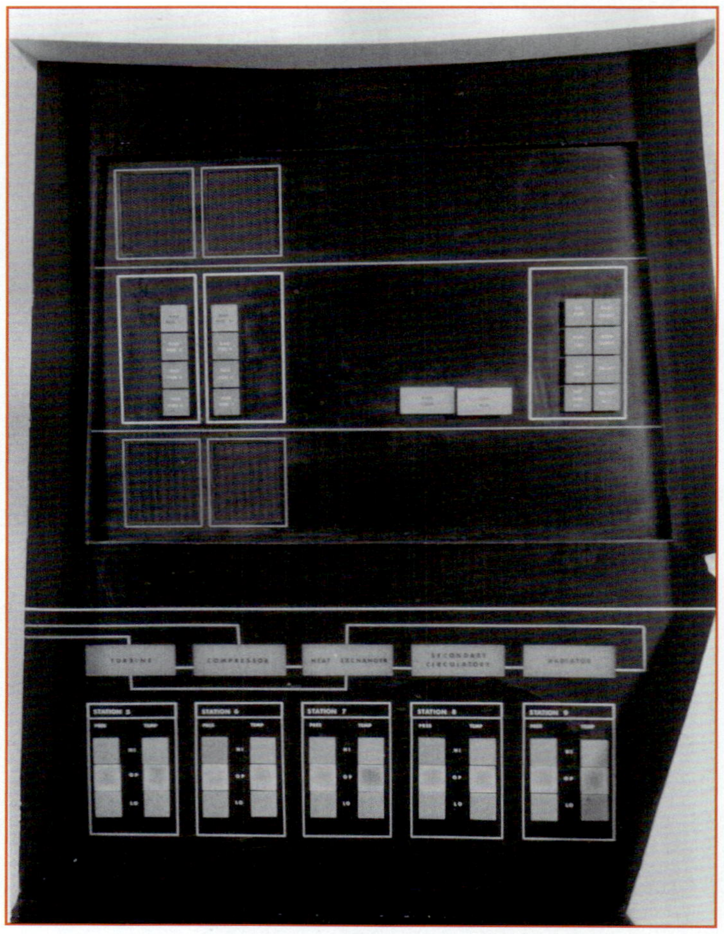

These two panels start, monitor and control the reactors' electrical output stage for power systems in the Discovery. Station 1 is the Centrifuge; 2, the Command Deck; 3, the Pod bay; 4, the EVA pods; 5, the 'Athena' room; 6 & 7, the storage and emergency entrance/exit; 8 & 9, the HAL computer.

A system flow chart is indicated by which each subsystem can be engaged, checked and monitored. Starting with the MHD generator and flowing through to the heat exchanger and finally the turbine and recuperator from which electricity is generated. The HAL 9000 can verbally alert the crew to any change or emergency condition. The small grilles have speakers and microphones to transmit or receive verbal information from the crew members, HAL, or EVA personnel.

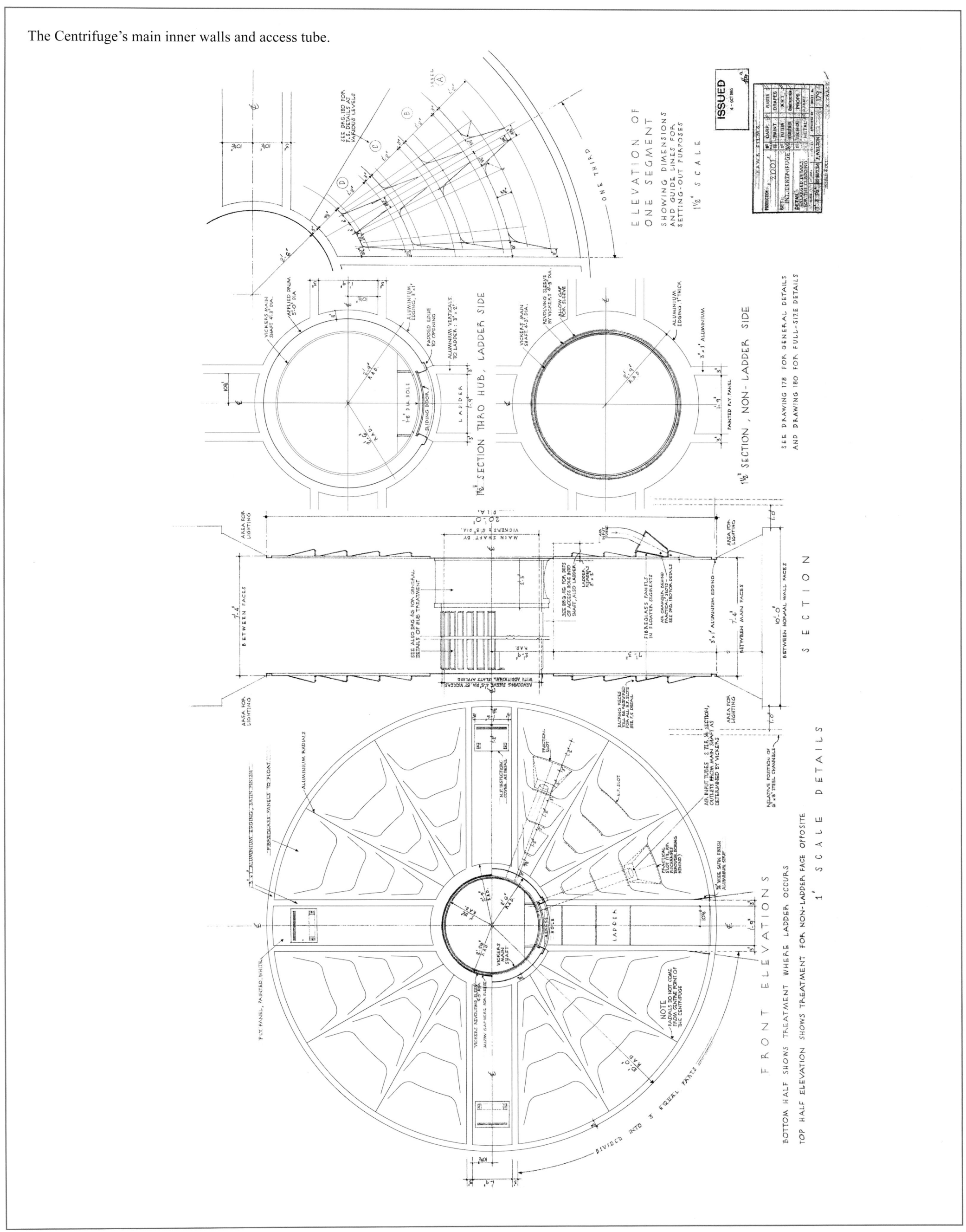

The Centrifuge's main inner walls and access tube.

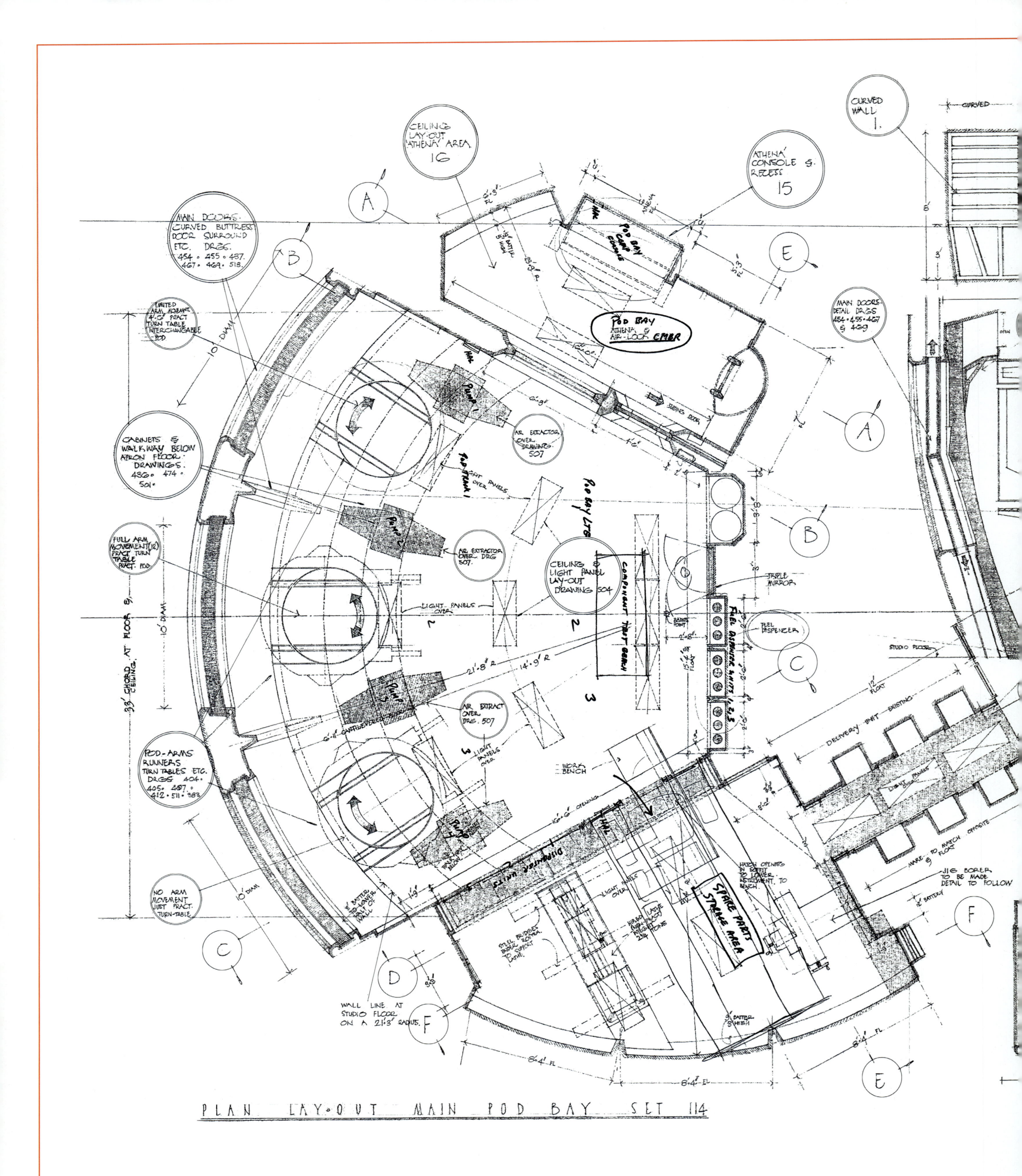

The Podbay

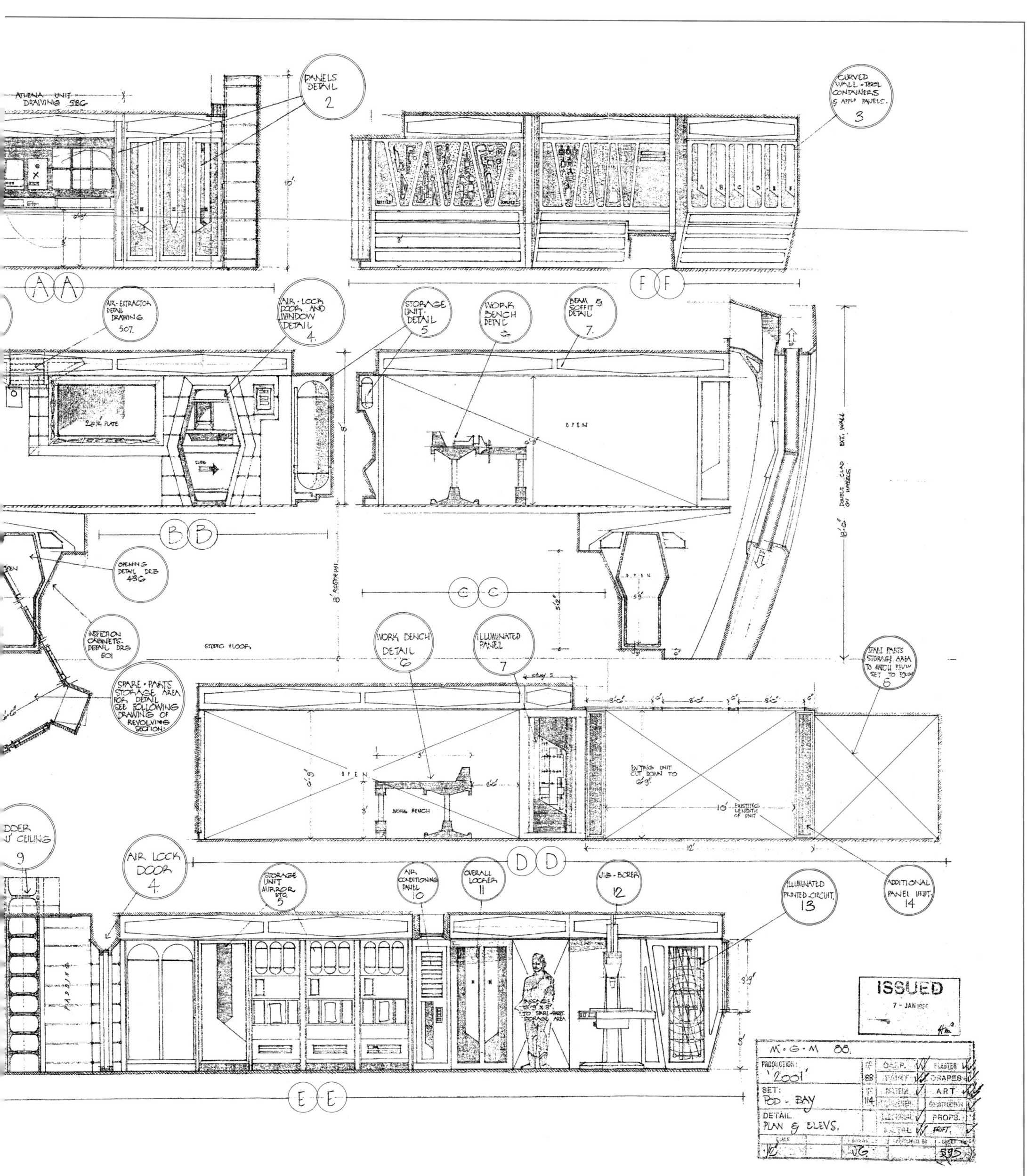

The master blueline drawing of the Podbay level layout. This drawing is part finalized and part work-in-progress. The areas marked F-F, D-D, and right third of E-E were never built. Manual changes are indicated by the handwritten marked areas. These areas were replaced by the rooms indicated on the next page.

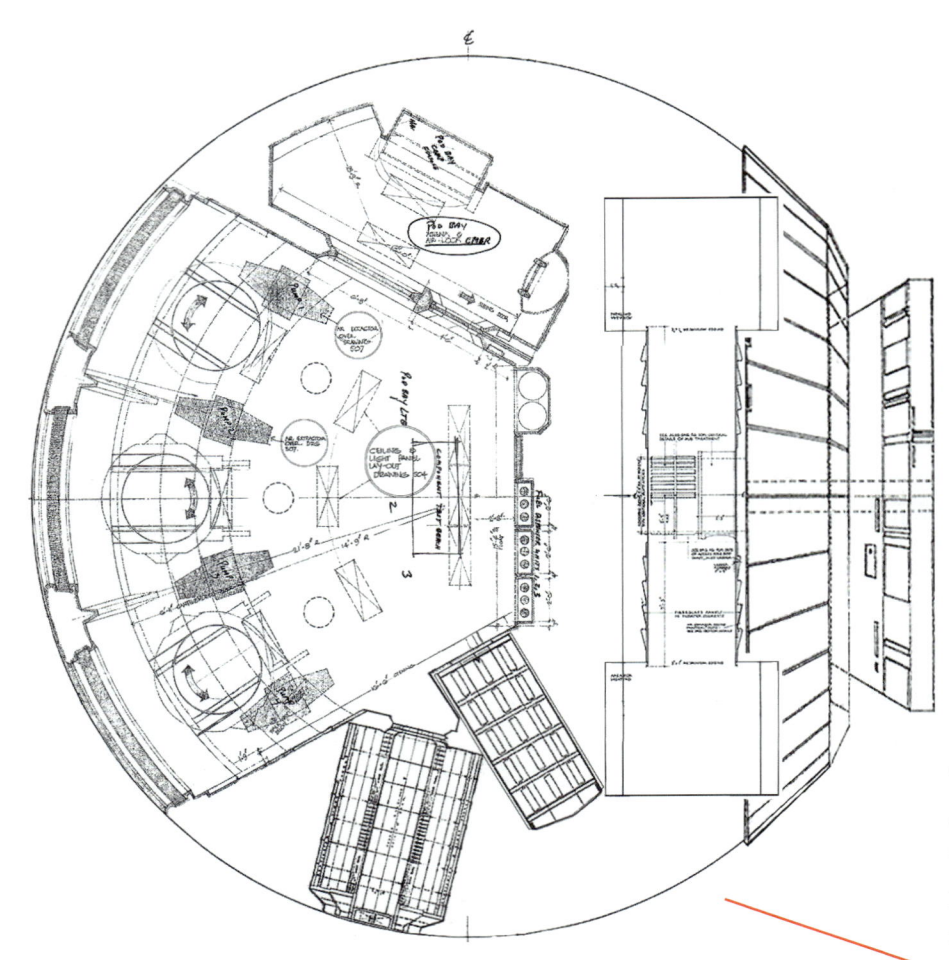

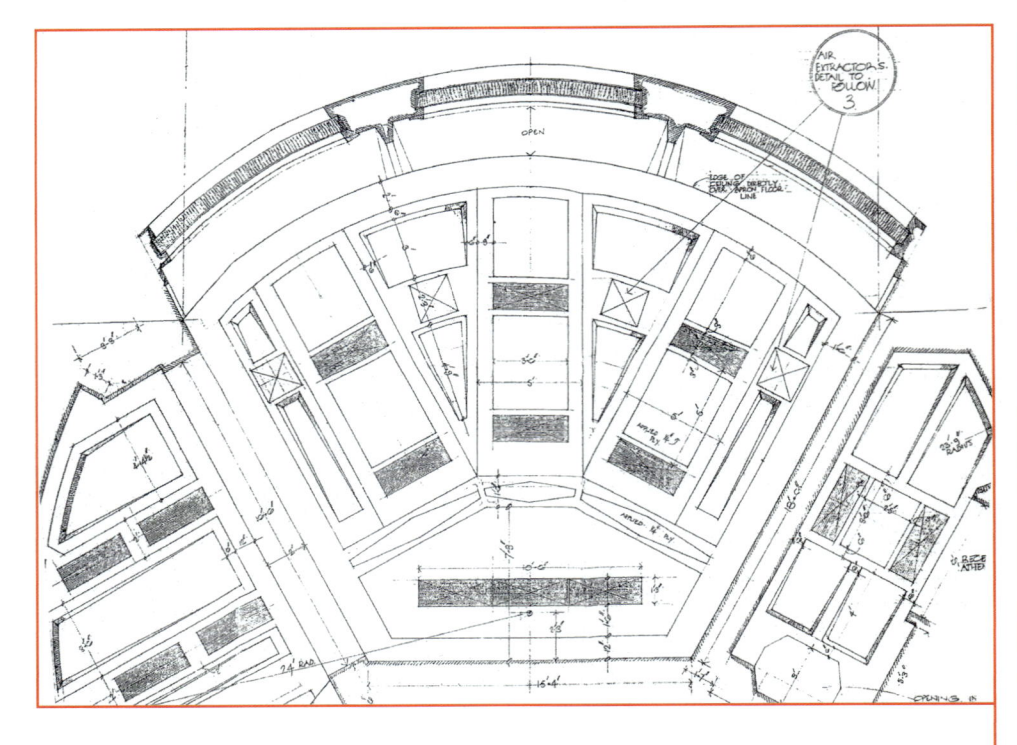

This plan view of the Podbay level of *Discovery* has each of the various blueline drawings superimposed and scaled to each other (accuracy +/- 2" to full scale) within the 52 ft diameter sphere. As with all space vehicles, everything tightly fits to maximize space. The Centrifuge has an internal diameter of 35 ft and maximum width of 10 ft. The Storage Area and Emergency Air-Lock were about 12 ft in internal length.

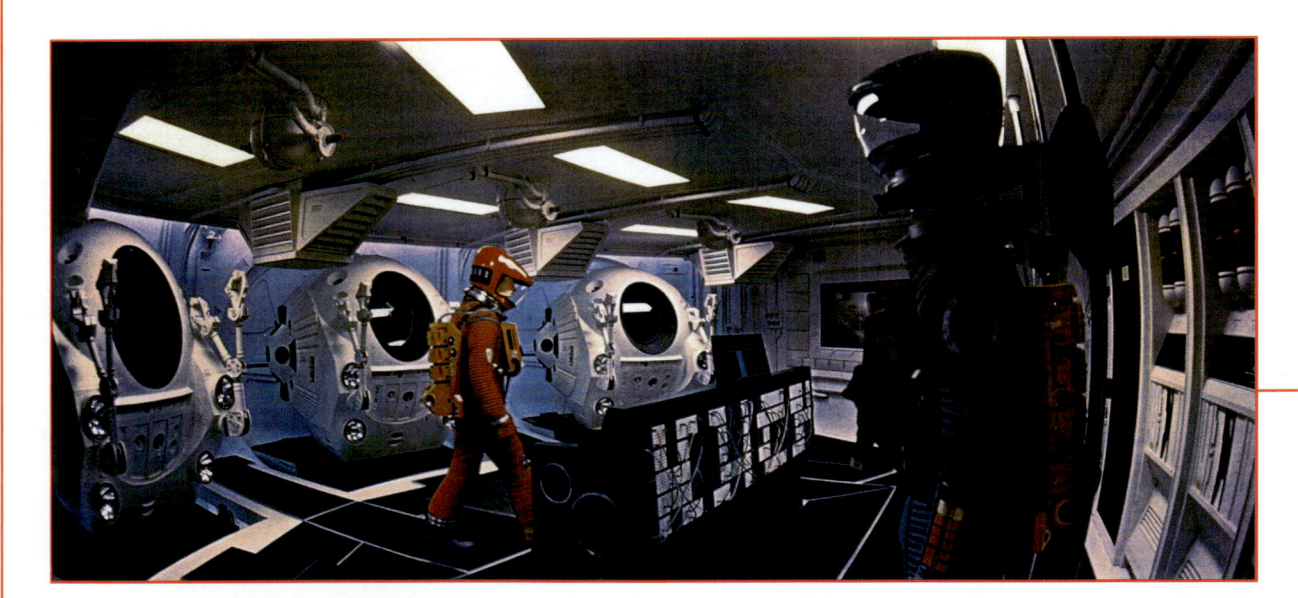

The ceiling plan of the Podbay and 'Athena' room looking down. Again, the area left of Podbay was replaced by the Emergency Entrance/Exit and Storage Area. Note: the Storage Area blueline drawing no longer exists.

View of the whole Podbay shot with a wide-angle fisheye lens. Note this picture has been artificially colorized (reasons unknown).

Back wall of Podbay view 1.

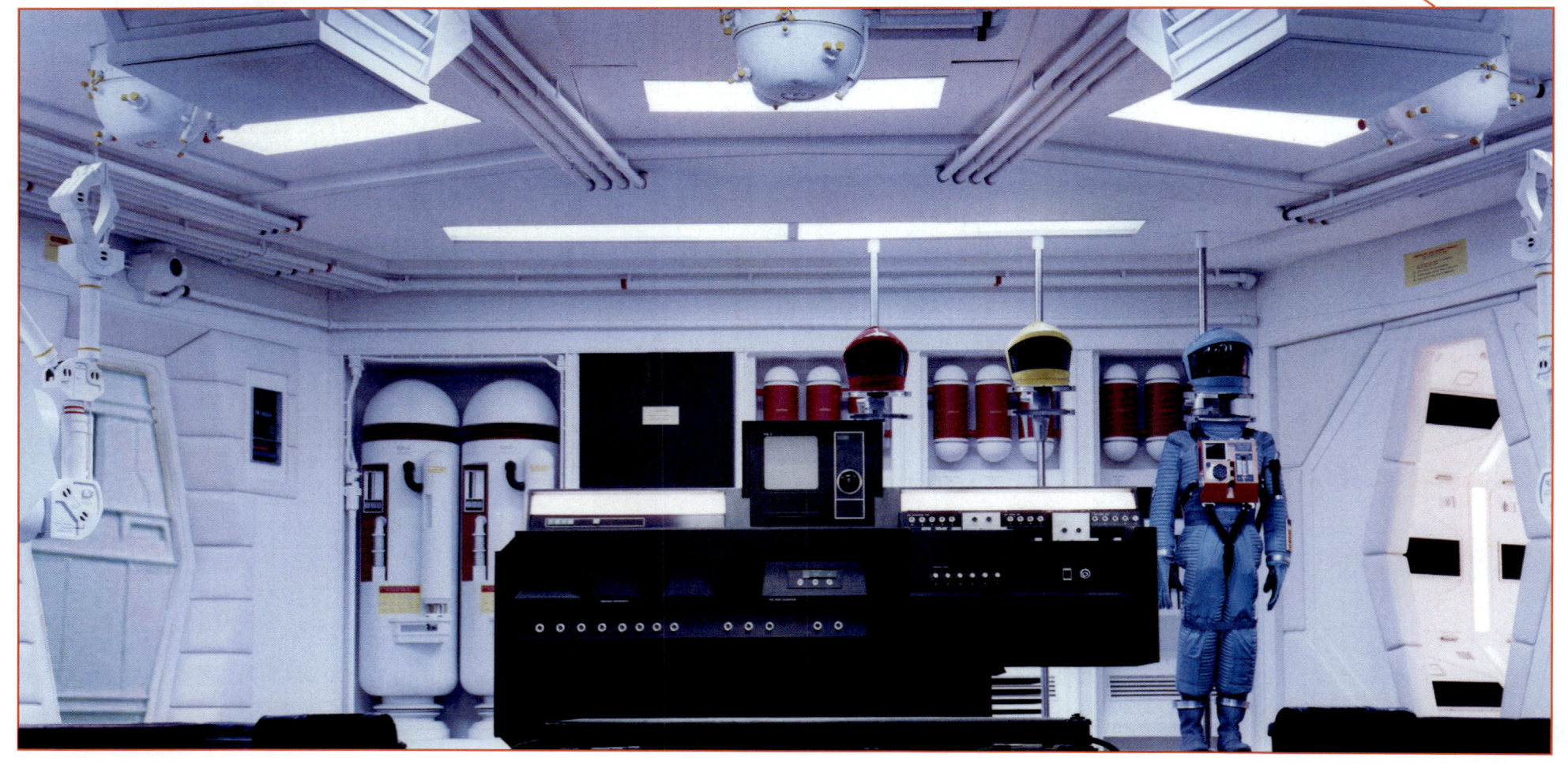

The rear wall displays EVA suits ready for wearing, and behind them are fuel cells for the pods. The lighted room in the background is the storage area.

These domed units in the ceiling run oxygen to the EVA suits, pods, and emergency pressurization via hoses that attach.

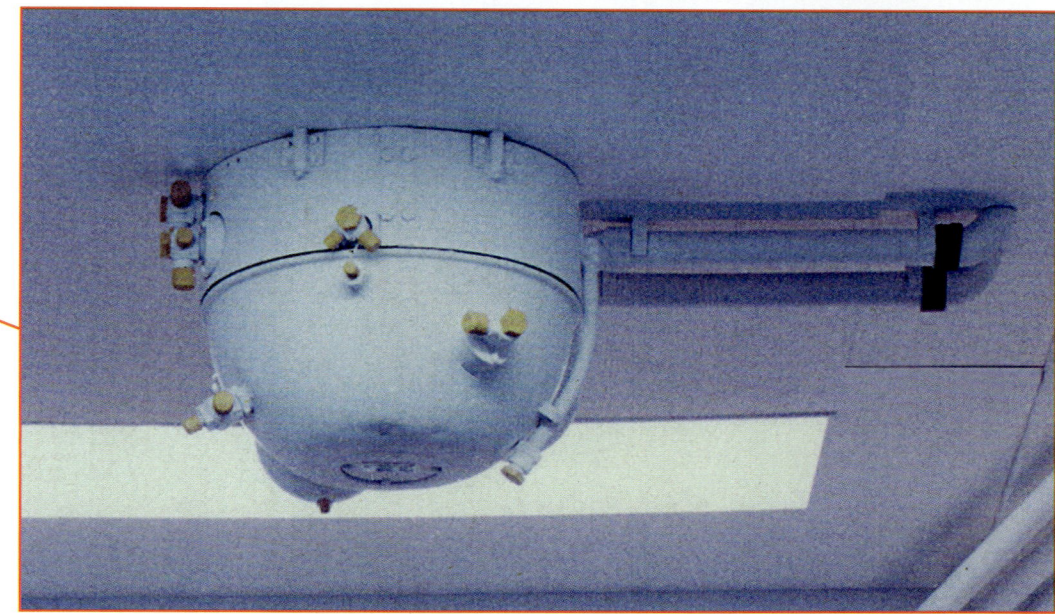

Two cylinders on back wall contain highly pressurized oxygen and nitrogen for repressurizing room when podbay doors are closed. Replacement cylinders found in the Tankage area at the spine of *Discovery*. A pod is sent out, the expended units are placed in the Tankage area, and new ones are transported back to the Podbay.

The Pods use three 'nitrodyne' fuel canisters that are inserted directly into the pods. Expended units are transported to the Tankage area.

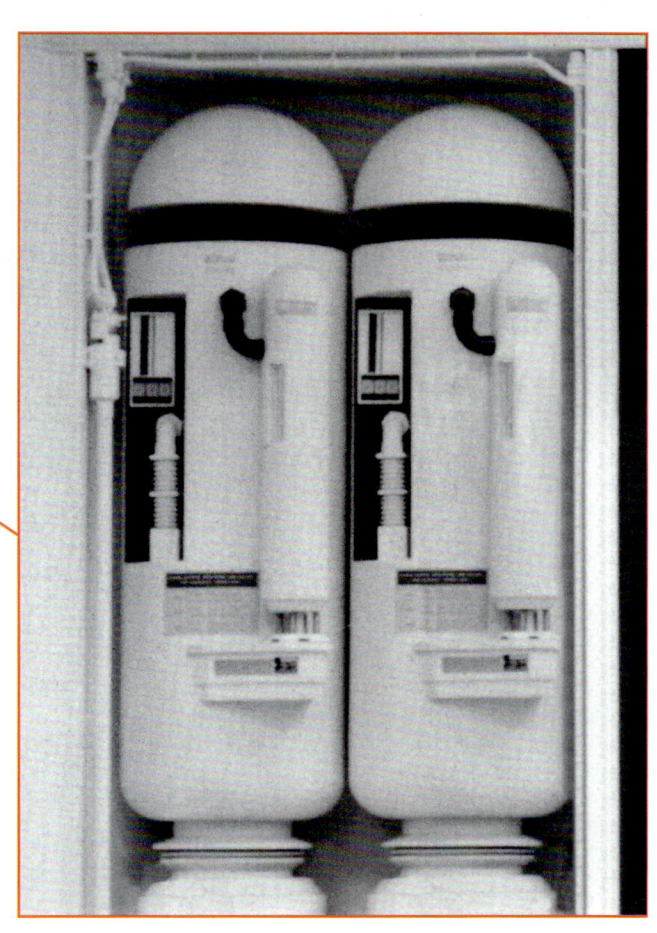

The Podbay Testbench

The test bench was built per drawing with the following exceptions:

Back elevation removable panels were replaced with a pegboard and ¼" jacks.

Second TV screen was added, and the HAL 9000 A/V interface rotated 90º to vertical.

Aluminum reeded areas replaced with fluorescent light panels.

Ordway and Lange check functionality

Rear patch bay area

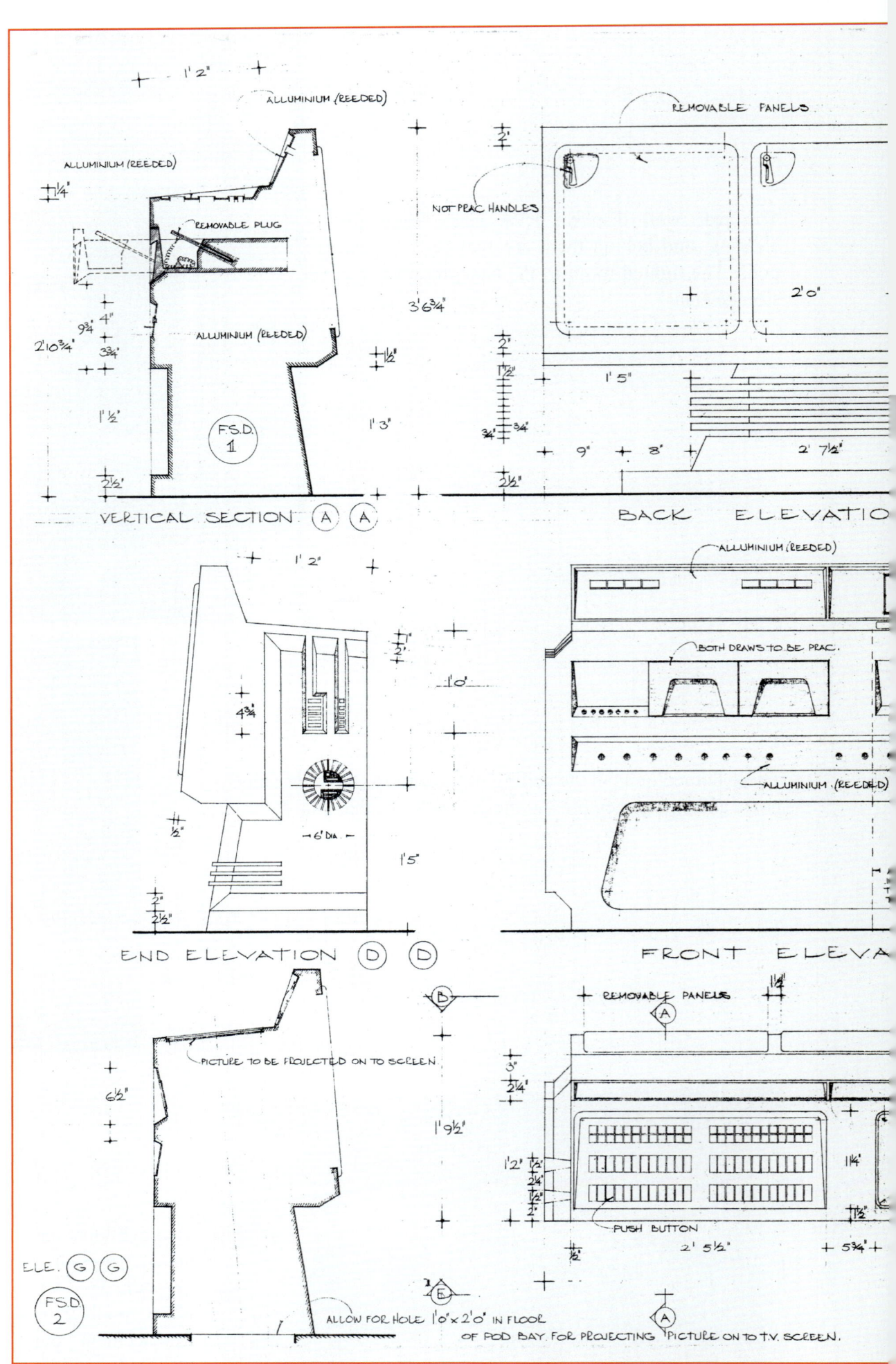

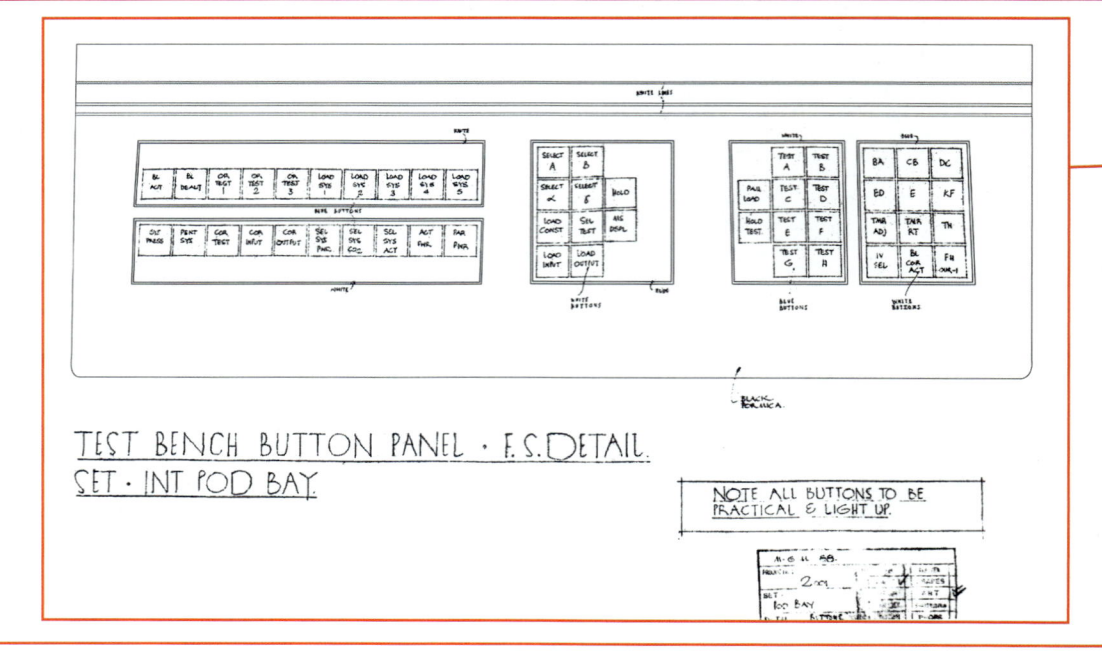

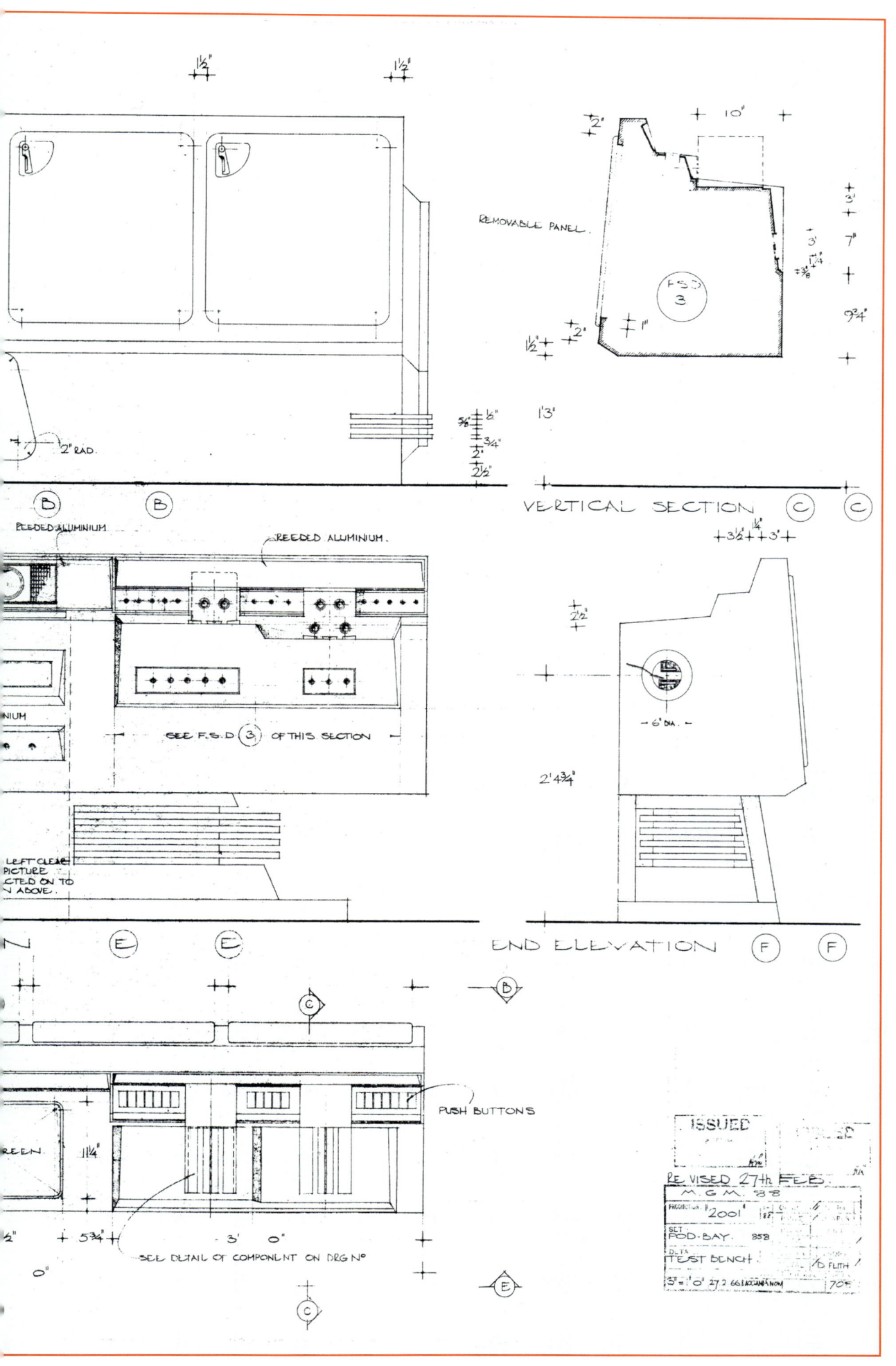

Right horizontal testing area

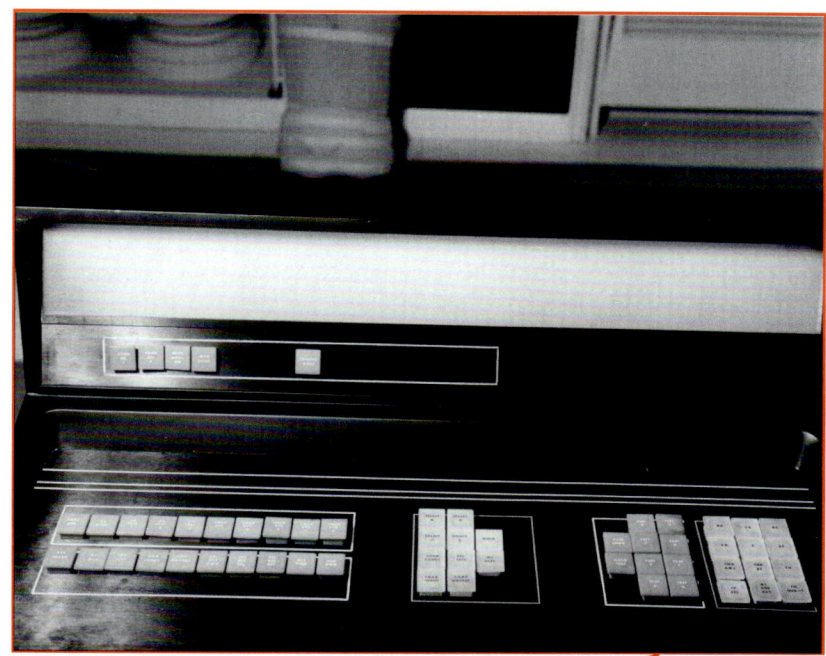

Left horizontal testng area

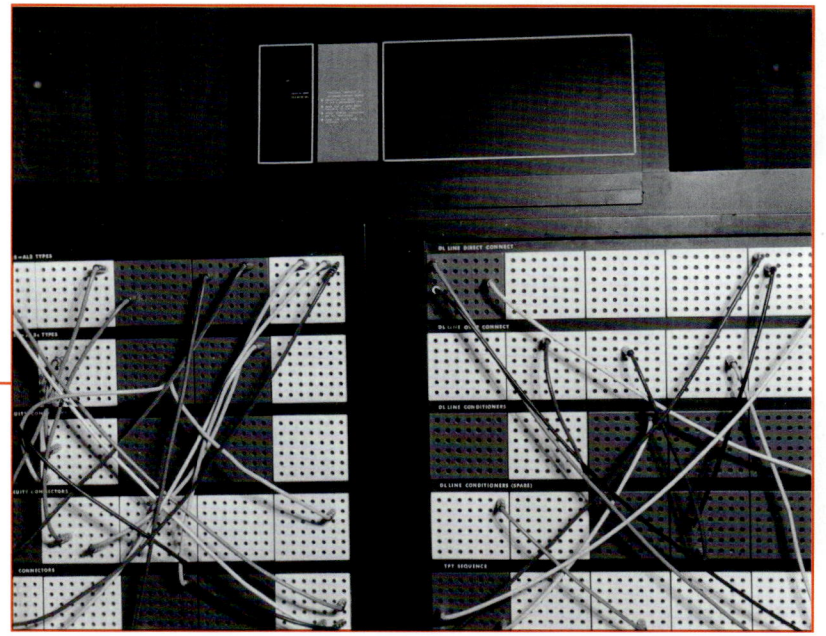

Revised upper left horizontal panel

Close-up of the test bench rear or 'Component Verification & Integrated Checkout Module'.

The EVA Pods

Space Pods

Three 7 ft diameter one-man space 'pods' were carried on board the *Discovery*. Although normally used to move around in space outside the spaceship for inspection, maintenance, and repair purposes, they had the propulsion capability to land on a small moon or asteroid and to shuttle back and forth from one space vehicle to another – say, in orbit around the Earth.

(a) Propulsion

A subliming solid system provides vernier propulsion, wherein the solid propellant sublimes at a constant pressure and is emitted from a nozzle. Such reaction jets will last for long periods of time, should have great reliability and use no mechanical valves. The main propulsion system is powered by storable liquids. This system, however, would only be employed in situations involving soft landing on a small moon or travelling at considerable distances from the *Discovery*.

(b) Mechanical Hand Controls

Selection controls were placed on each side so that the appropriate hand must be removed from the manipulator to select a tool or to park. Selection of a tool returns the arm to the 'park' position, where it leaves the 'hand,' then the arm goes to the appropriate tool and plugs in. In doing so, it inhibits the 'finger' controls on the manipulator so that when the operator returns his hand into the glove he can only move a solid object, not individual fingers.

(c) Television

It was found possible to produce all-round TV coverage with eight fixed cameras. This, however, did not give a sufficiently accurate picture for docking or selecting a landing space. For this purpose, it is assumed that the field of view can be narrowed and orientated; controls were included for this purpose.

Normally the TV link would be occupied by the internal camera, so that the parent craft can monitor the pod interior. The pod pilot can switch in any other camera for specific purposes (survey, etc.) reverting to interior camera for normal work.

(d) Proximity Detector

This is the safety system with omni-directional coverage working from the main communication aerials. It gives audible warning when the pod approaches a solid object. This is necessary as a safety measure as the pilot cannot monitor seven or eight TV displays continuously. The system also detects an approach to an object, the speed of which is too high to be counteracted by the vernier thrust settings on the control system. In this event, full reverse thrust is applied, over-riding the manual control setting. The system depends upon a frequency modulated transmission and under safe conditions was assumed to result in a low, soft audible background-signal. This continuous signal was considered necessary in order to provide a continuous check on a vital safety system. If the speed of approach to an object became dangerous compared with the distance from it, the tone would become louder and higher pitched and, if unchecked, end in a shrill note accompanied by reverse thrust. The system could also work in conjunction with a transponder (to give the necessary increased range) to measure distance from *Discovery*.

(e) Flying Controls

Manual controls were considered necessary both as a stand-by and for local maneuvers. Two hand control sticks, each with two degrees of freedom and fitted with twist grips, provide the necessary control about six axes.

Analogue information is presented for attitude, heading rate and distance; these can be referred to local ground (for landing, take-off, etc.), course (which enables the pilot to face forward, head up, on any pre-selected course.), or parent ship (for docking, local maneuvers, etc.). These data had to be presented, as the pilot had to act immediately on them. This is the most easily assimilated display. A variation in full scale rate, which can be applied by the control sticks, is included; this allows the full stick movement to result in any proportion of full vernier motor thrust, thus giving a 'fine' control for local maneuvers.

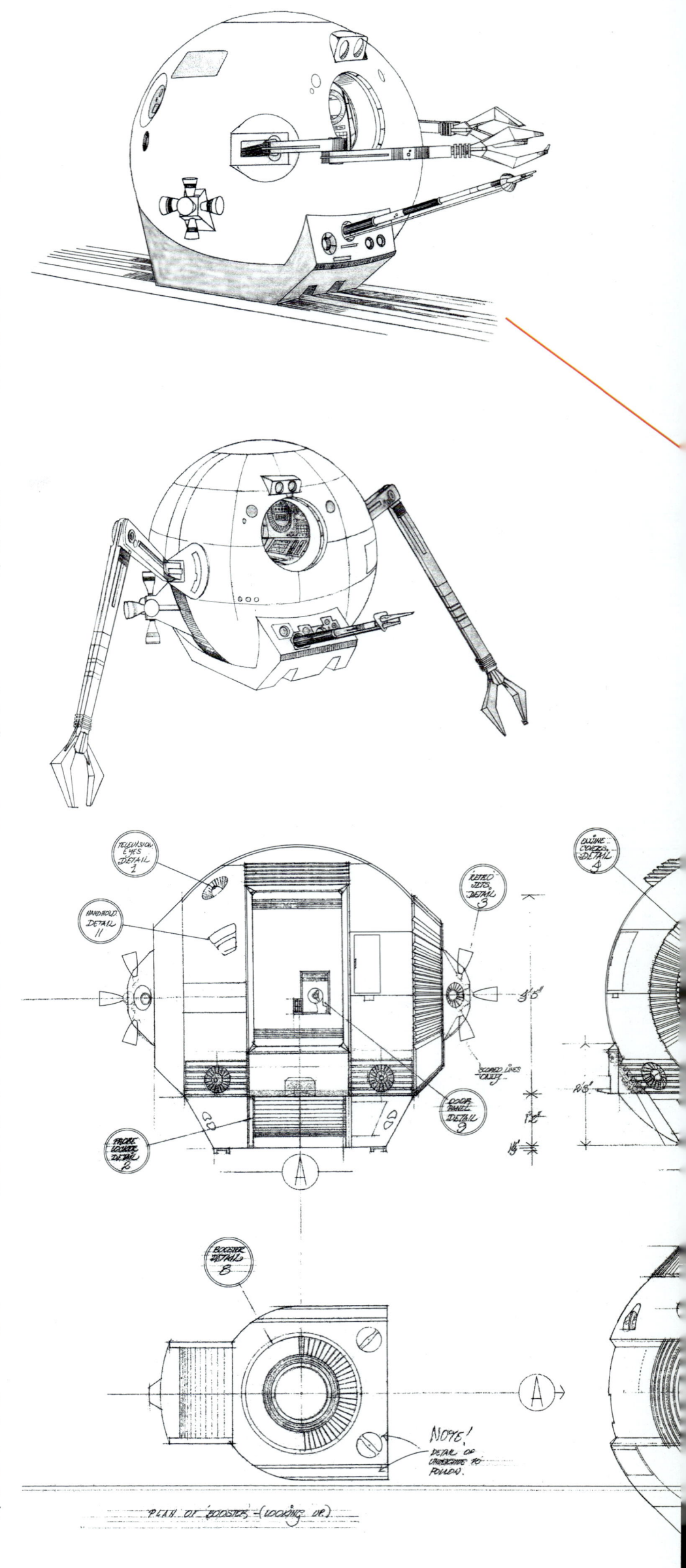

The three pods in the Podbay. Picture taken with a wide-angle semi-fisheye lens.

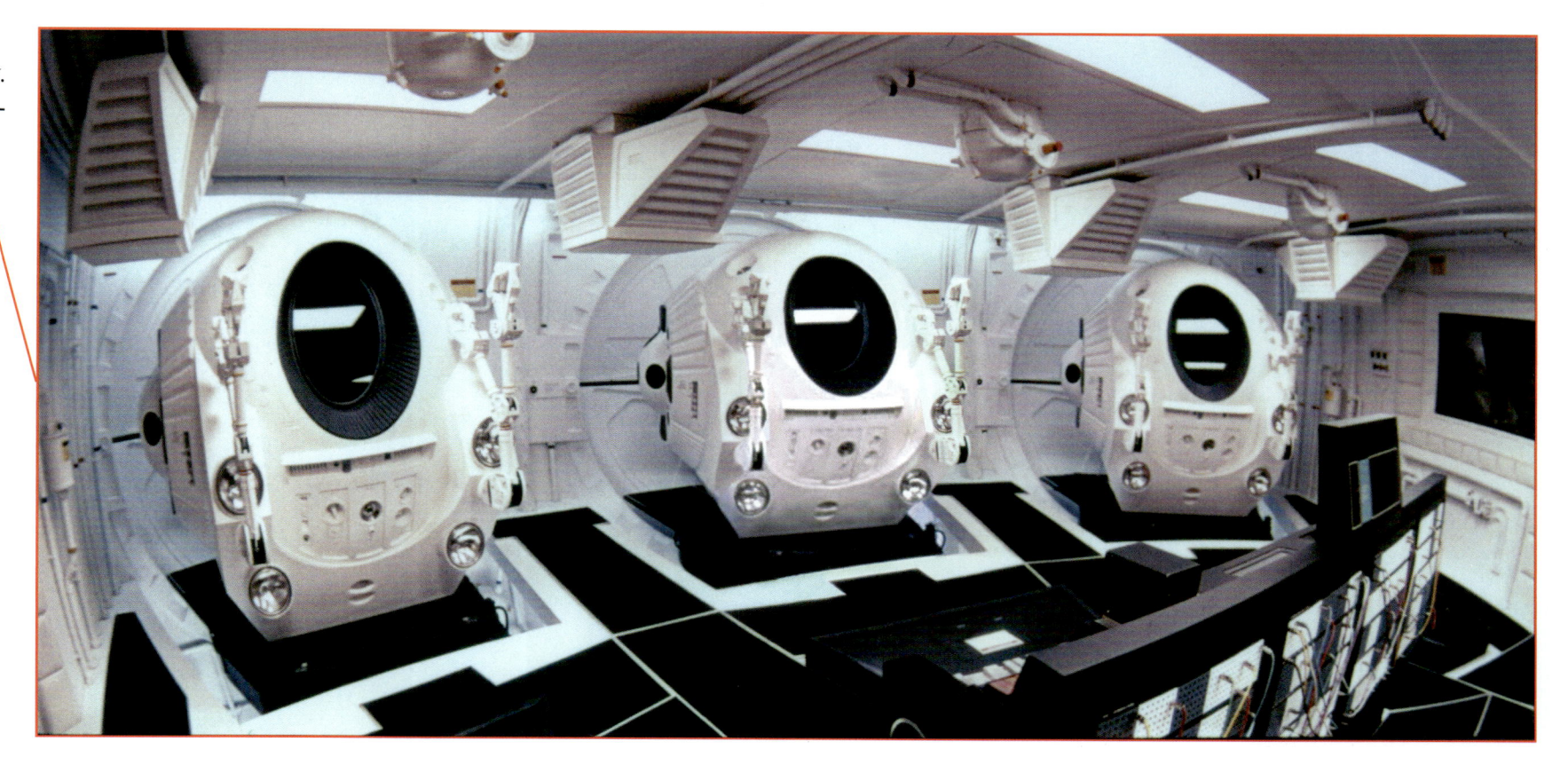

This concept sketch by Harry Lange shows the evolution of the design from a basic 'pressure sphere' design. Tremendous thought was placed into functionality.

Finalized drawing of the EVA Pod. Note the arm and hand controls were relocated (per photos) and redesigned. Four headlights were also added.

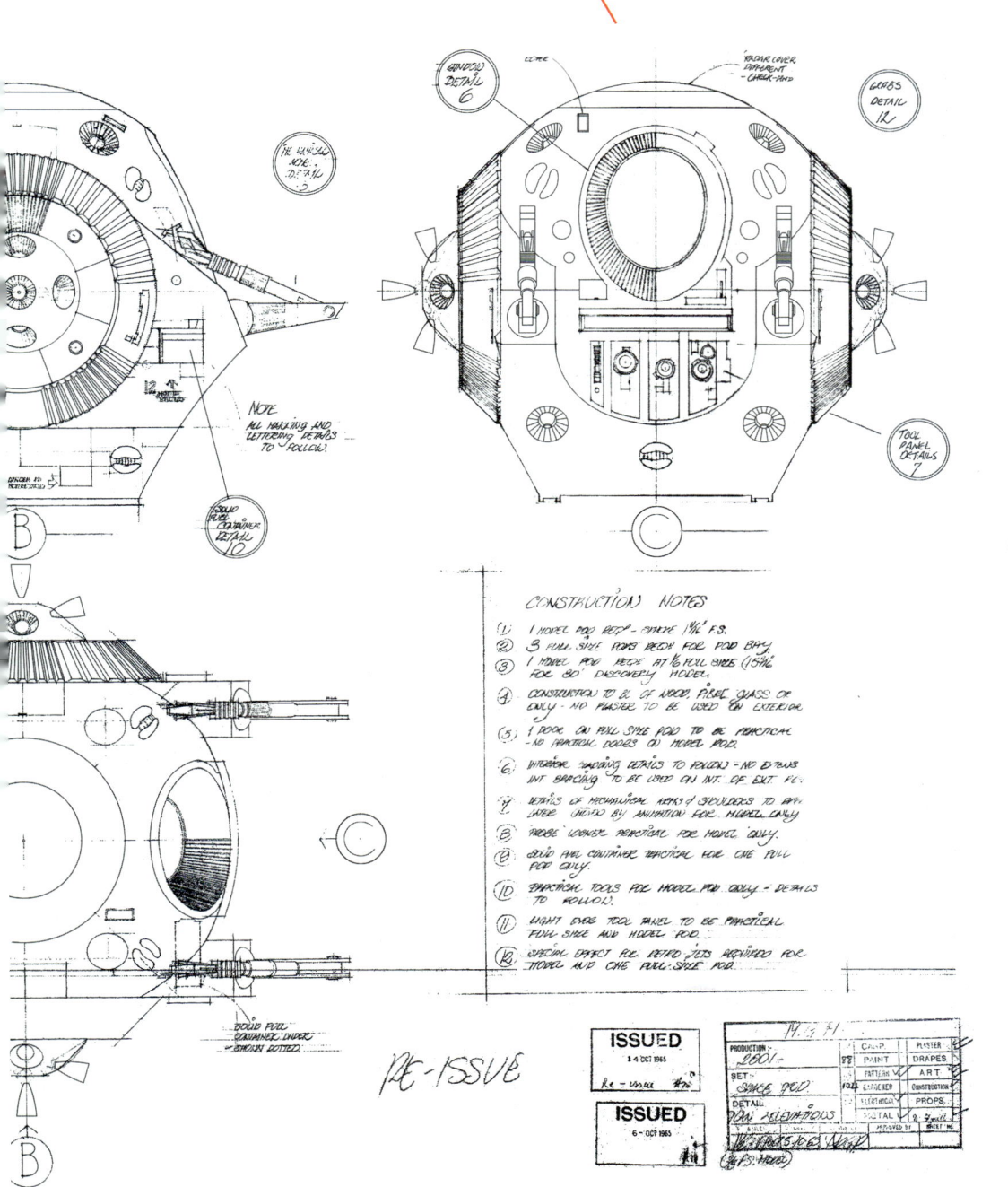

Space pioneering writers Ken Gatland (left) and Les Sheperd (right) with Ordway, Lynn Sheperd and son.

Ordway and Lange demonstrate the EVA Pod's functionality.

71

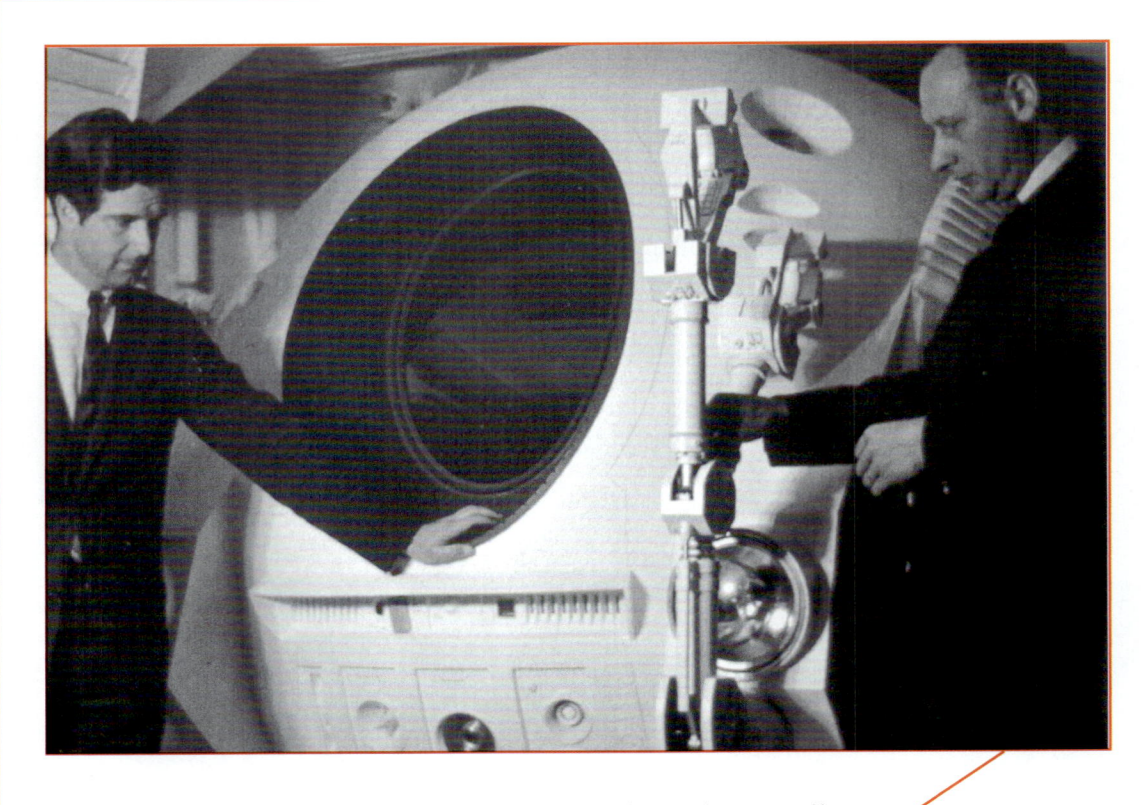

Ordway and Lange demonstrate the 'Arm and Hand' controllers.

Rear of EVA pod exiting the Podbay

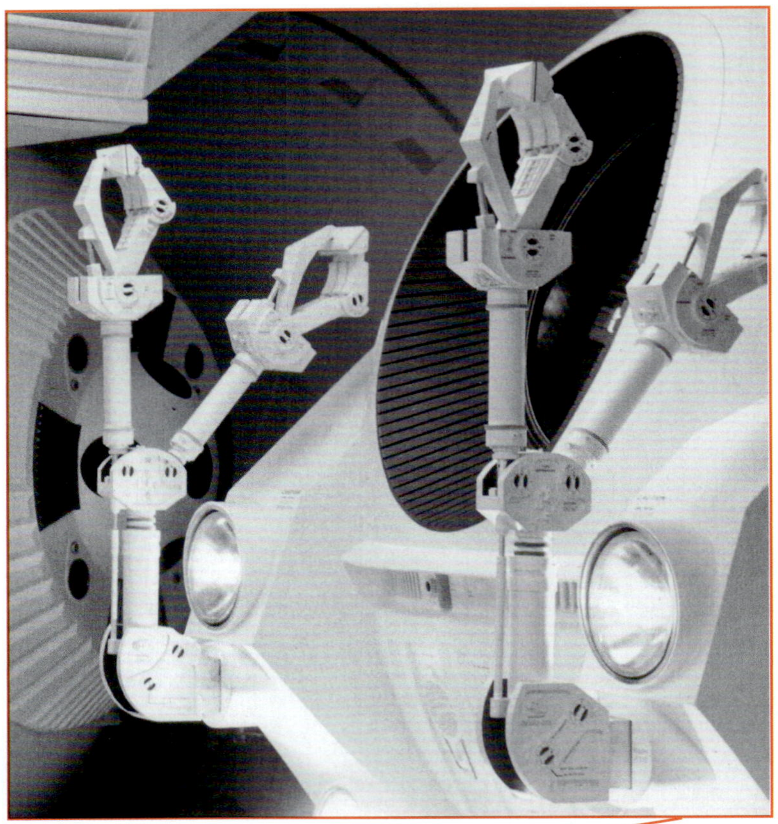

Close-up of Pod arms and hands

Close-up of Pod rear entry door

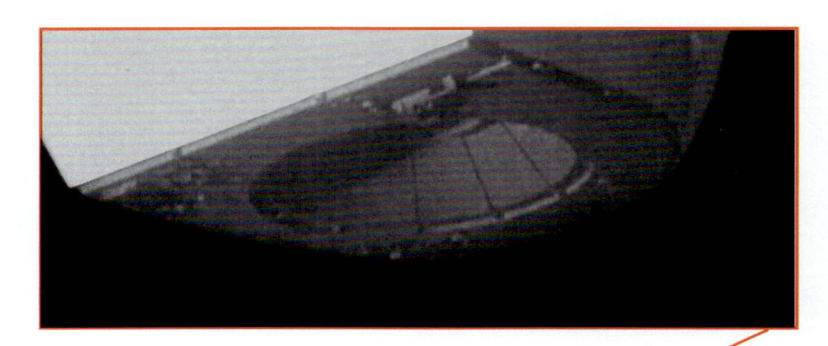

Close-up of vertical launch booster on underside

In April 1966, a Russian attache of scientists visited the Discovery spacecraft. *Left to right*: Cracknell, Boris Polikarpov, Anatoly Chuev, Ordway, Clarke, Kubrick, Unsworth.

Several months later, this Russian EVA pod was designed for the Soviet moon program.

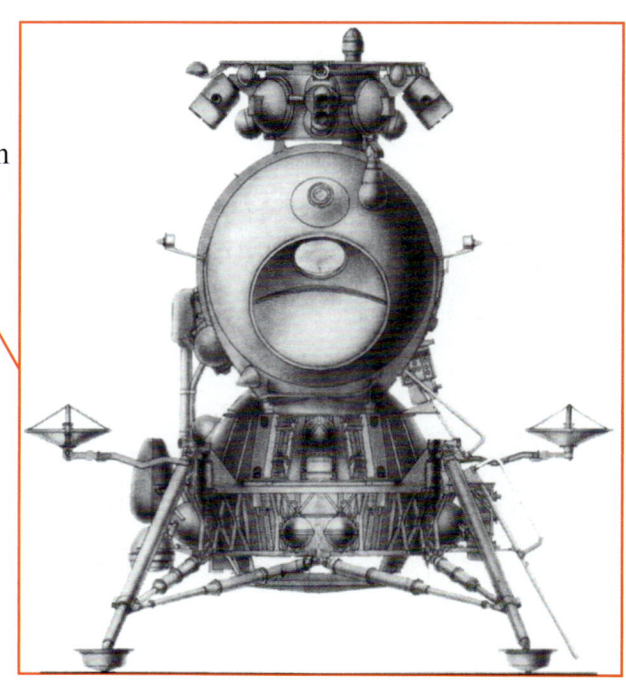

EVA Pod Interior Details

The emergency release system can blow the rear hatch off when the astronaut needs to vacate promptly.

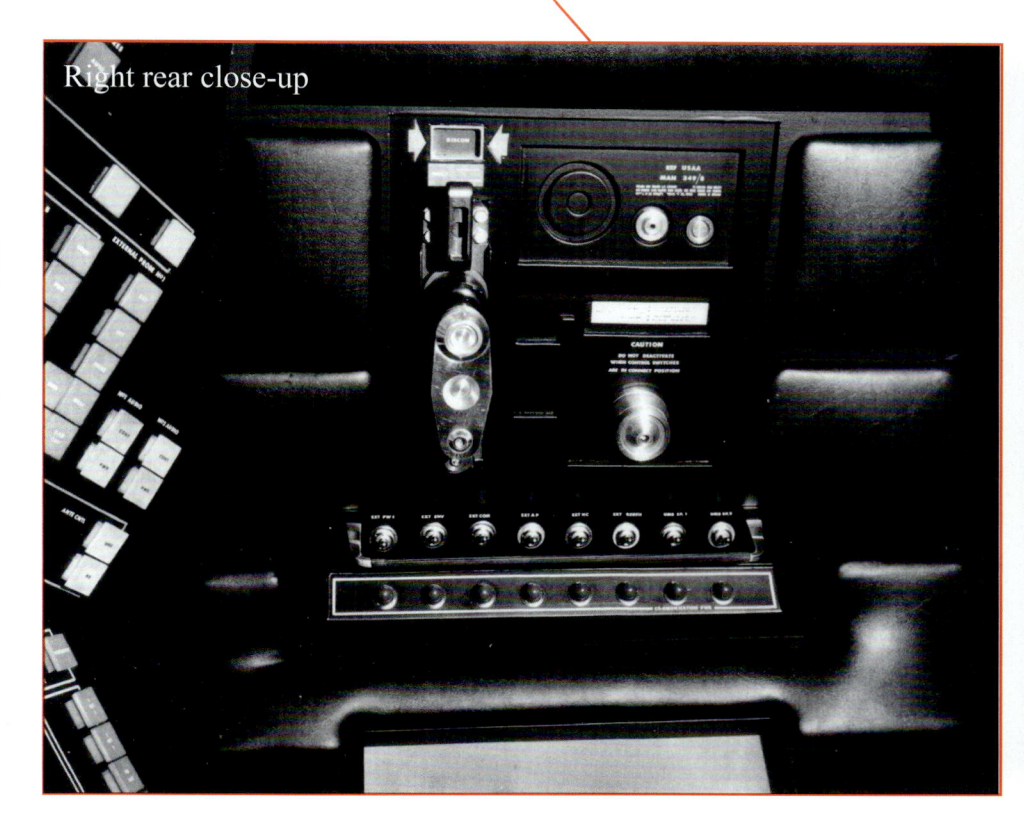

Right rear close-up

Right rear view

Two video monitors are utilized via an outbound camera on the front of the pod and another on a jettisoned probe. The buttons fine-tune and align the cameras manually.

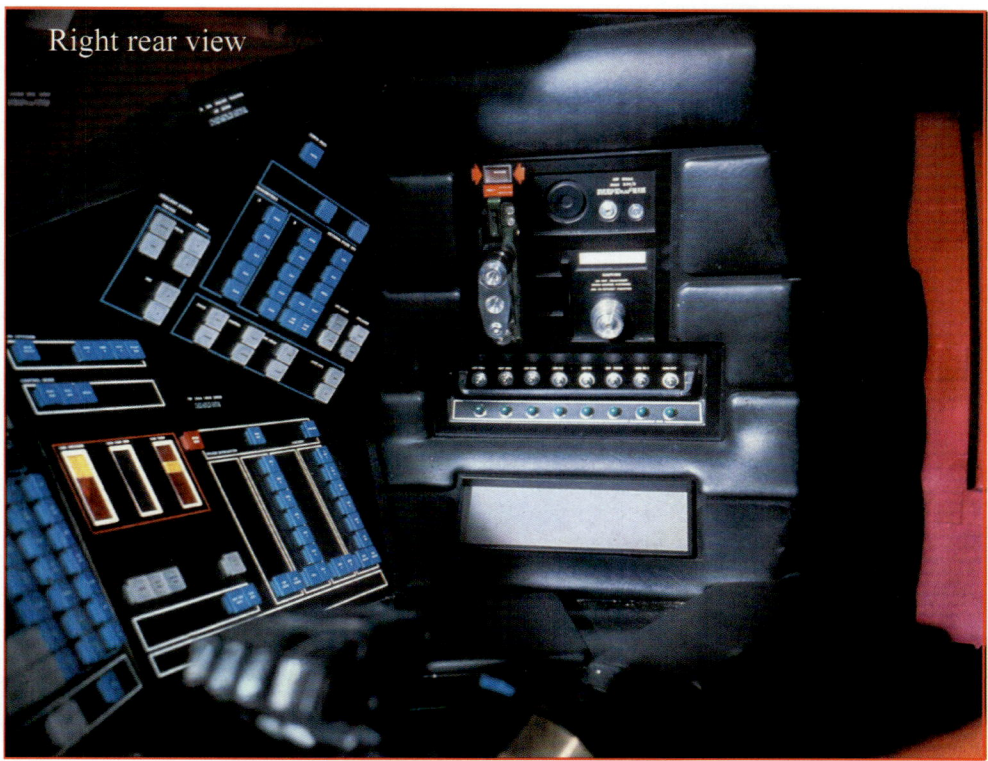

These panels check power distribution, cabin pressure, radiation and temperature. A probe (mounted in the front recessed panel between the headlights) can be jettisoned and monitored.

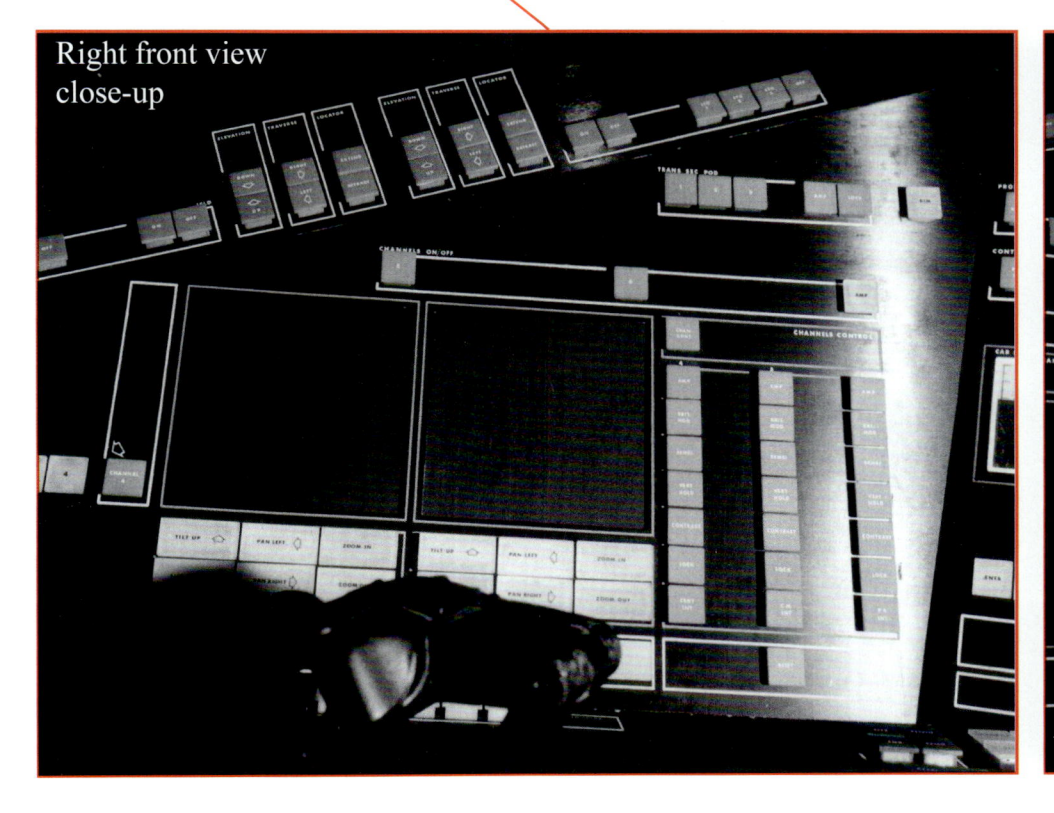

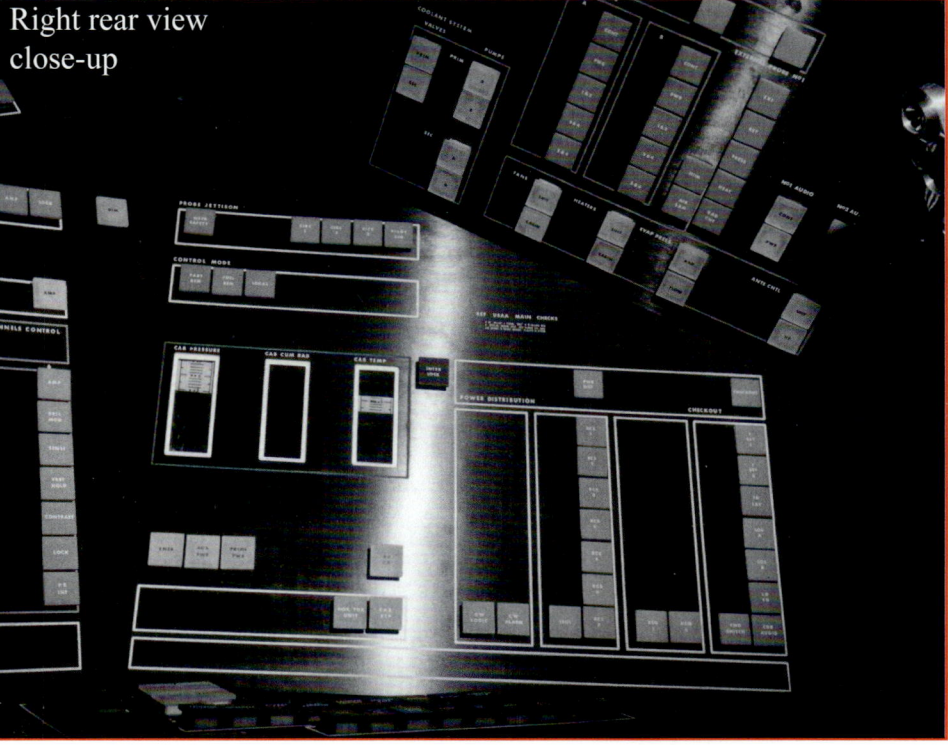

Right front view close-up

Right rear view close-up

73

Center panel displays fuel consumption information. These two hand-grips manually control the large 'Manipulator' arms and claws on the front of the pod. Rotating right moves the arm accordingly, pulling back-or-out, slides the arms into stowed position (for example).

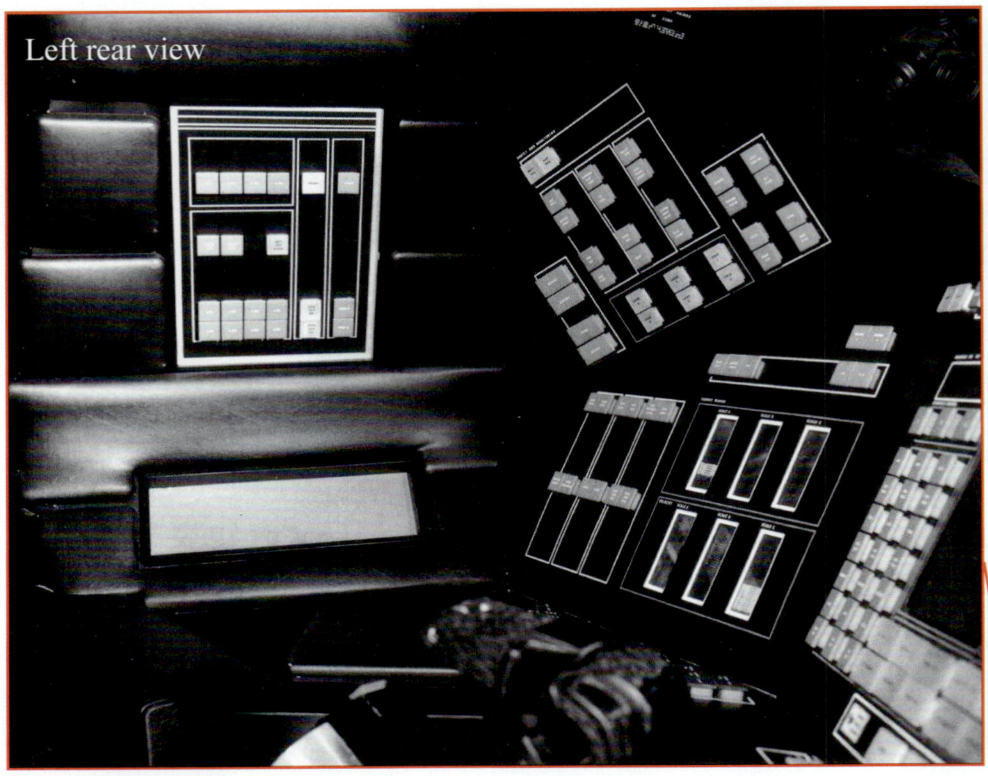

Left rear view

Center lower panel

These panels control a forward firing laser that returns information of the surroundings (geophysical). Information can be stored as well. A sensor on an arm can be projected out of the front of the pod and is monitored and controlled here.

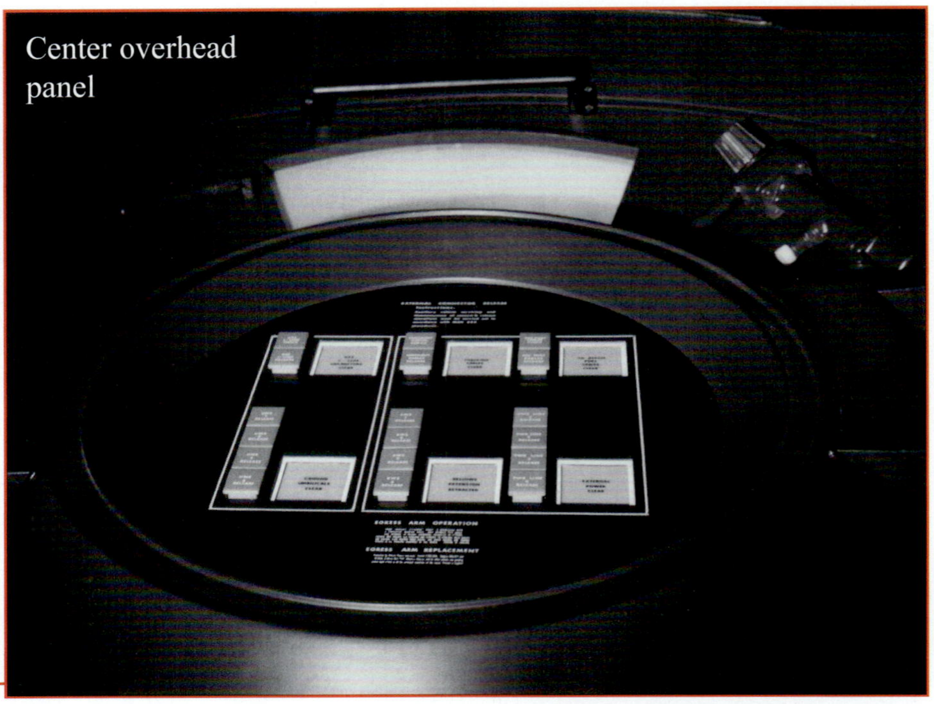

This round overhead panel controls the safe release of the Pod from the Pod platform that slides out of the Podbay (Egress arm) via connection release monitors. Six lights activate alerting the astronaut of checkout cables, umbilicals, the access port and connectors that have been cleared. The air pressure bellows valve then disengages and the Egress arm retracts into the Podbay. Ultra-sonic Motion Sensors (UMS) allow the astronaut to know if the pod gets too close to an object, and the Safety Warning System (SWS) engages with an alarm.

Center overhead panel

These two TV monitors display information from the laser scans and the forward sensor. Also displays HAL 9000 random interface monitoring with pod.

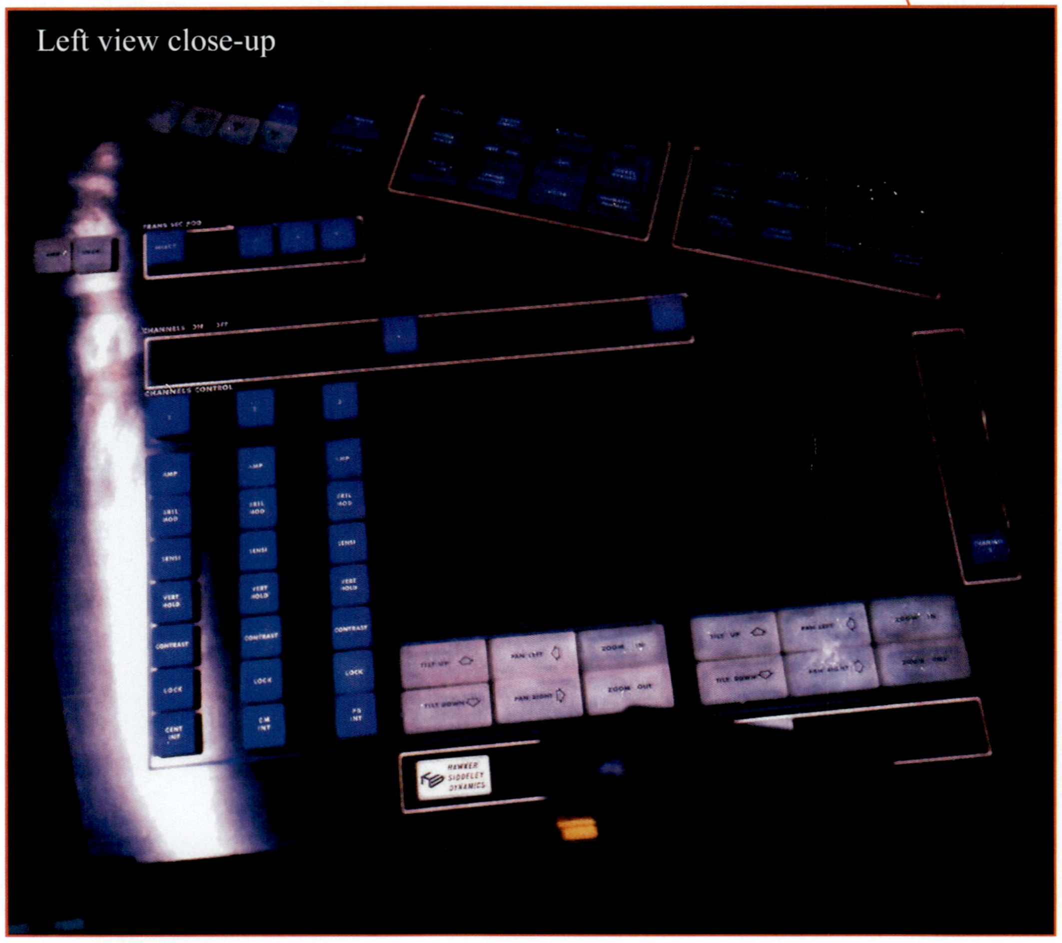

Fred Ordway inspecting the inside of the Pod.

Left view close-up

74

The Athena Room

This room is the starboard side of the Podbay. It contains a back-up HAL 9000 computer and dedicated scientific analysis station. The upper command level is accessed via a ladder through this room.

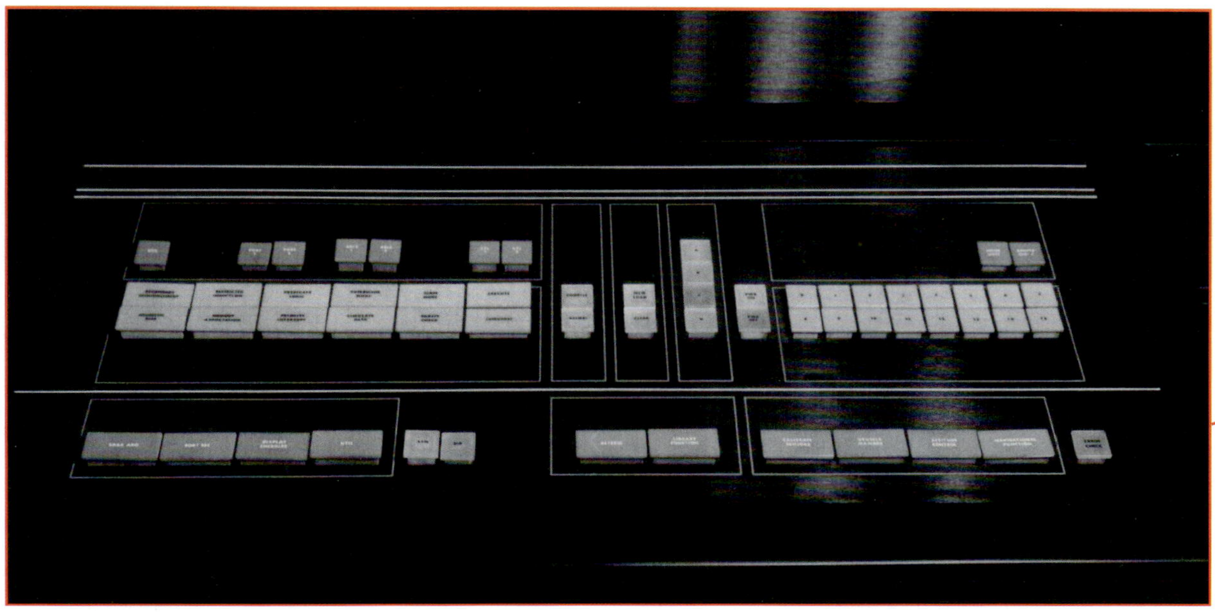

This console (directly under the six video displays) is a user digital interface between the main HAL9000 computer and the Pods. The Pod can be interrogated before, during or after a mission to view and store information gathered. All systems in the Pod are checked and calibrated from here.

This console is for the 'Digital Packs' (on the back of the EVA helmets) to be downloaded, programmed and or viewed (after an EVA mission for instance). A library function stores all individual EVA missions in detail for later view or retrieval. The console also allows a special software program to be created, developed and formatted to a specific function.

The Equipment Storage Room

View looking into storage room from podbay.

View looking out of storage room into Podbay. An astronaut locates a new electronics unit for maintenance or replacement of various critical systems in *Discovery*.

The inside storage room of the International Space Station has a similar design, with all four surfaces used for storage. Space inside habitat areas are utilized to maximum efficiency. (Photo: Courtesy NASA)

The Discovery EVA Suit

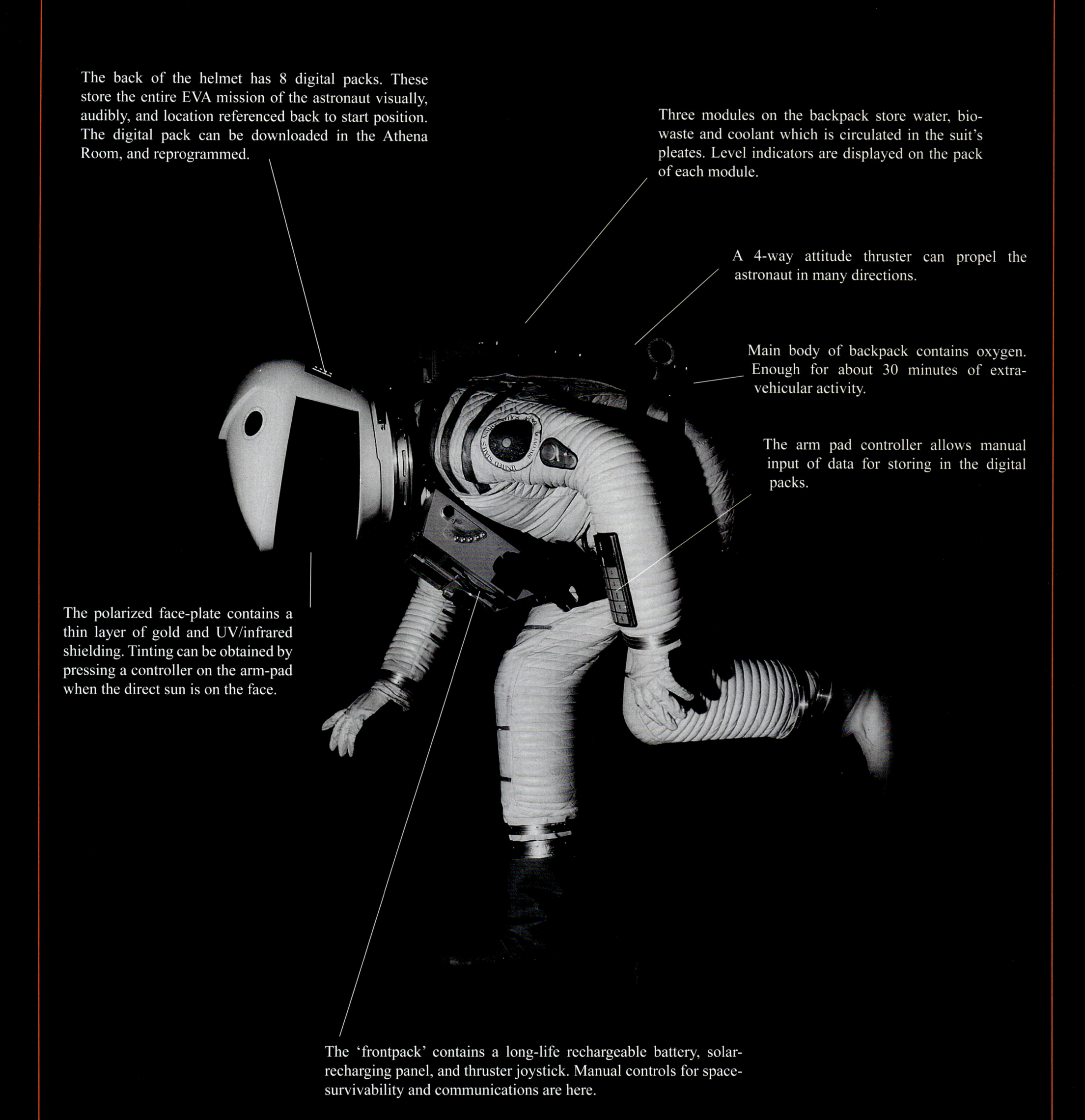

The back of the helmet has 8 digital packs. These store the entire EVA mission of the astronaut visually, audibly, and location referenced back to start position. The digital pack can be downloaded in the Athena Room, and reprogrammed.

Three modules on the backpack store water, bio-waste and coolant which is circulated in the suit's pleates. Level indicators are displayed on the pack of each module.

A 4-way attitude thruster can propel the astronaut in many directions.

Main body of backpack contains oxygen. Enough for about 30 minutes of extra-vehicular activity.

The arm pad controller allows manual input of data for storing in the digital packs.

The polarized face-plate contains a thin layer of gold and UV/infrared shielding. Tinting can be obtained by pressing a controller on the arm-pad when the direct sun is on the face.

The 'frontpack' contains a long-life rechargeable battery, solar-recharging panel, and thruster joystick. Manual controls for space-survivability and communications are here.

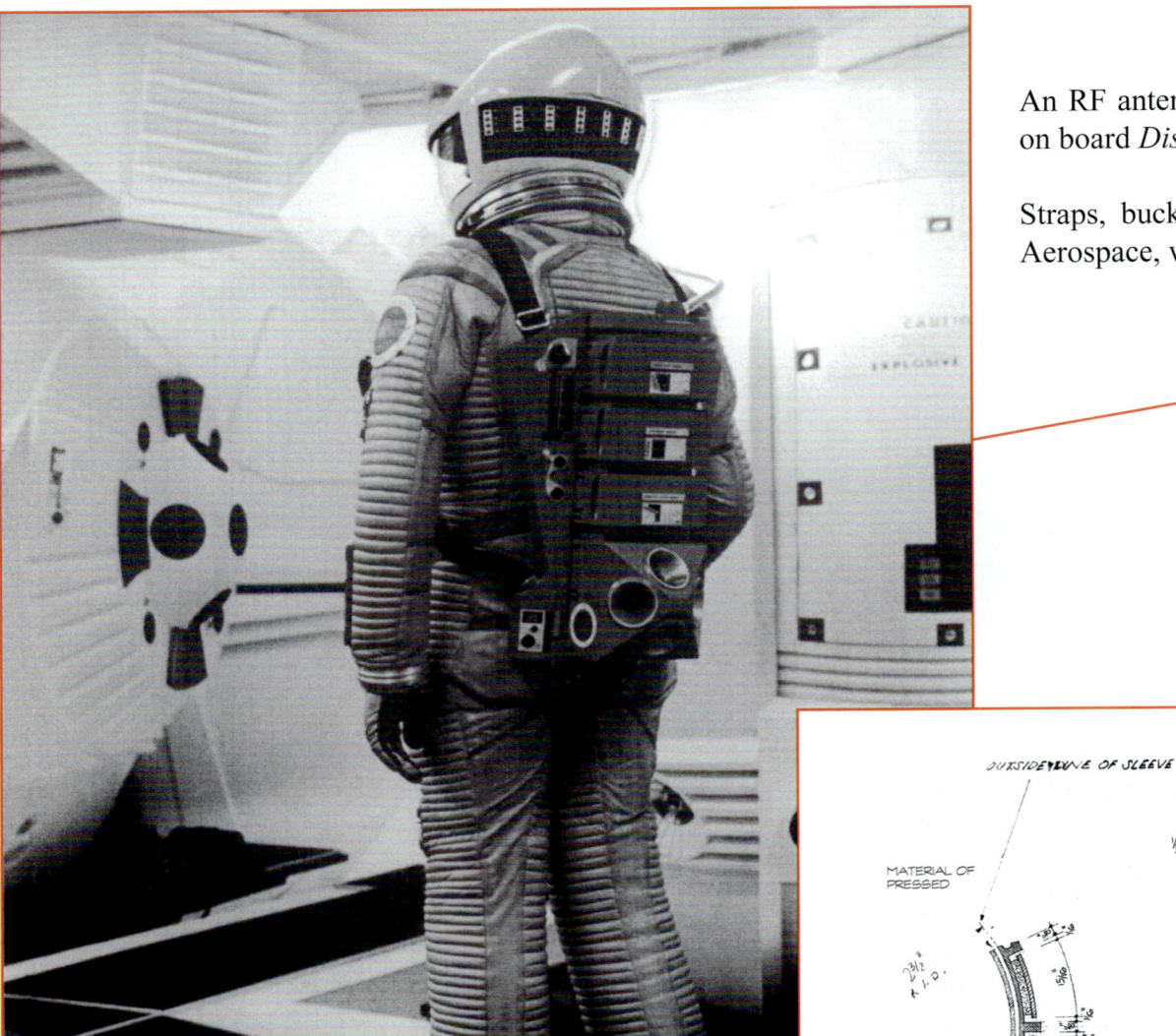

An RF antenna maintains oral communications with HAL or personnel on board *Discovery*.

Straps, buckles, and other fitting hardware supplied by Martin-Baker Aerospace, well-known for supplying the West with ejection seats.

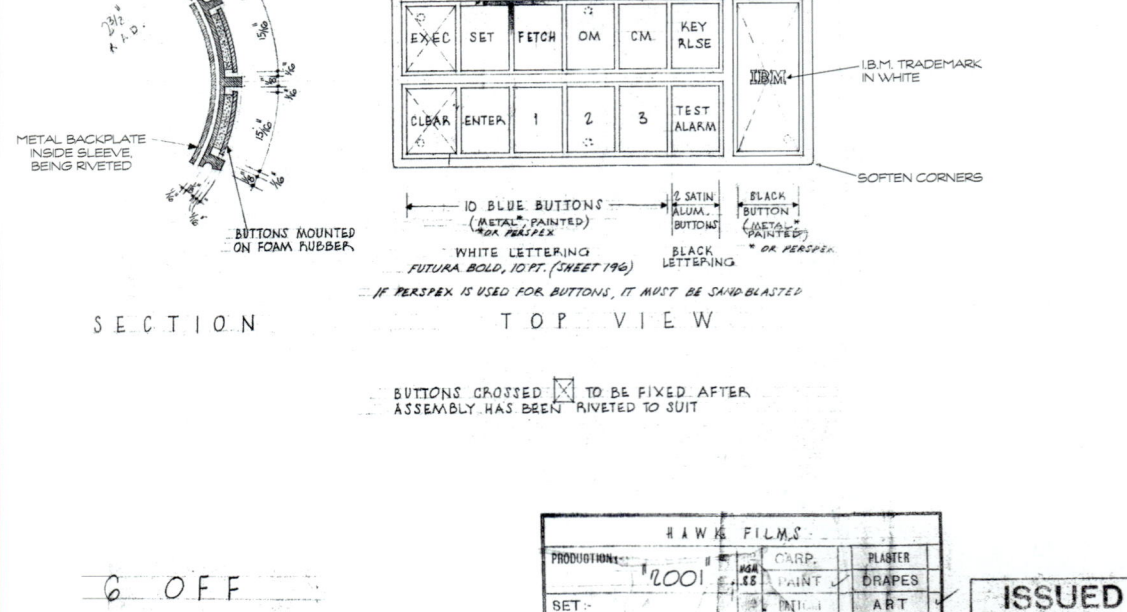

Blue-line drawing of the Arm Controller, The Final hardware is a mirror image of the schematic drawing.

Four EVA suits are on *Discovery:*

Red – mission commander

Yellow – deputy commander

Blue – Astrophysics

Green – Geophysics

EVA Frontpack Environmental Control Unit (ECU)

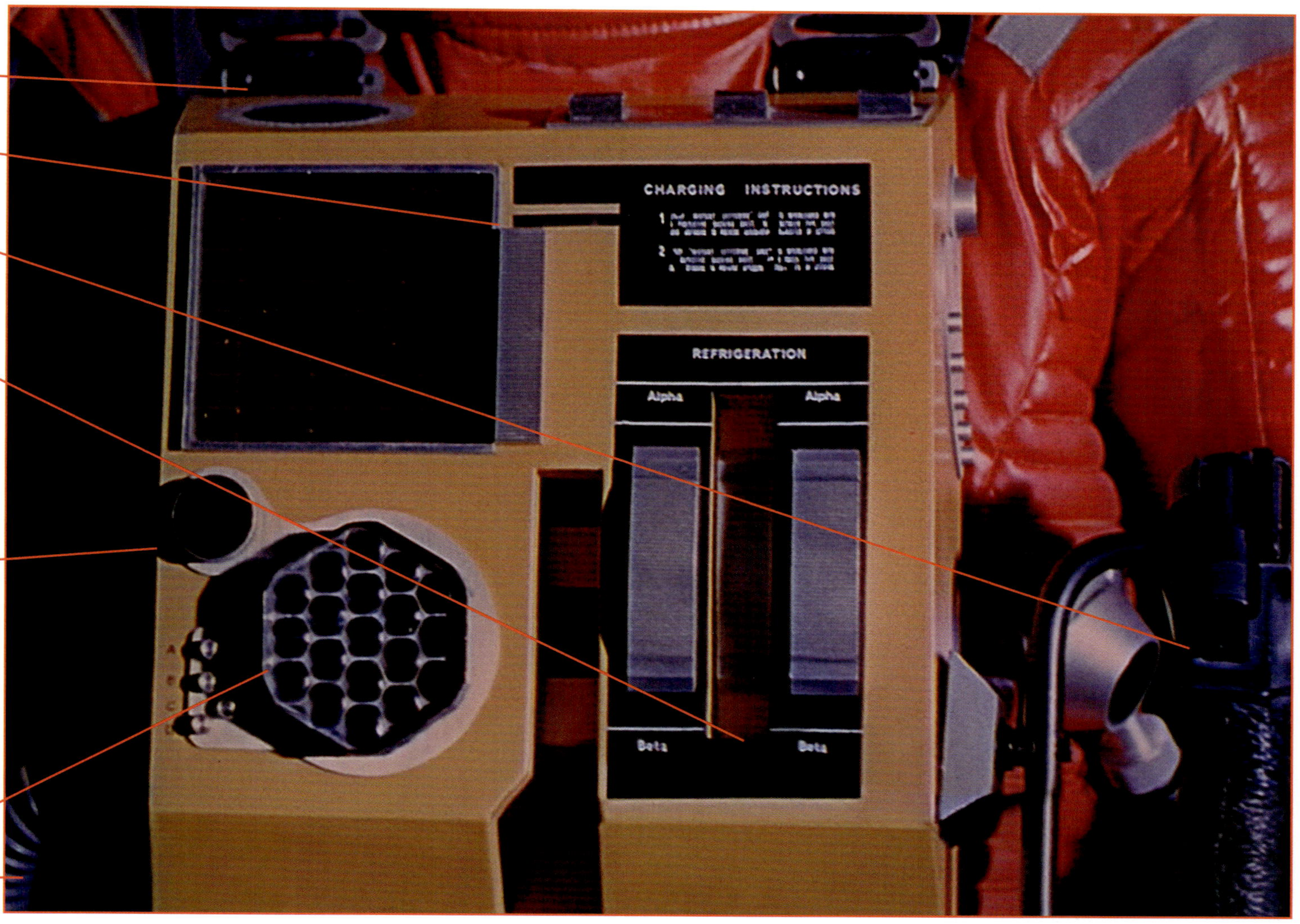

3-way locking belt clamps provided by Martin-Baker.

Switch activates screen.

Thruster joystick supplied by DeHavilland Aircraft.

Alpha refrigeration warms astronaut's skin while drawing perspiration away. Beta refrigeration cools astronaut skin while drawing perspiration away.

Switch can be manually selected for parts of suit for solar battery recharging:

A. Thrusters
B. Refrigeration
C. Digital Systems
D. Environmental

Solar collection cells recharge battery.

Refrigerant return hose.

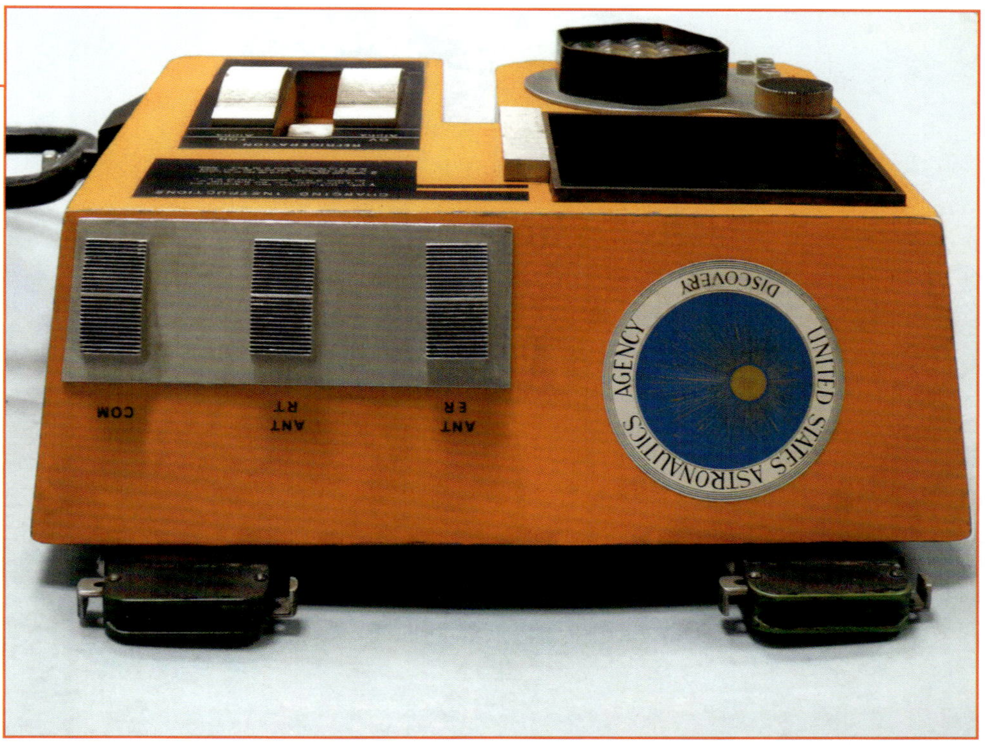

Top view of ECU pack

Three switches on top of the pack engage communications via the antenna on back of helmet.

ER ANT: Extended Range Antenna permits only audio communications. Useable past 100 feet.

RT ANT: Optimized Antenna carrying range of 800-2100 MHz transmits video and audio communications.

COM: engages communications with *Discovery*.

Port view of ECU pack

The joystick gives the astronaut fine maneuverability control of the 4-way pulse thrusters on the backpack. The black round knob ties the umbilical from astronaut to the Pod or *Discovery*. Lights A through E indicate activation of thrusters.

The Emergency Air-Lock

This room is located portside of Podbay.

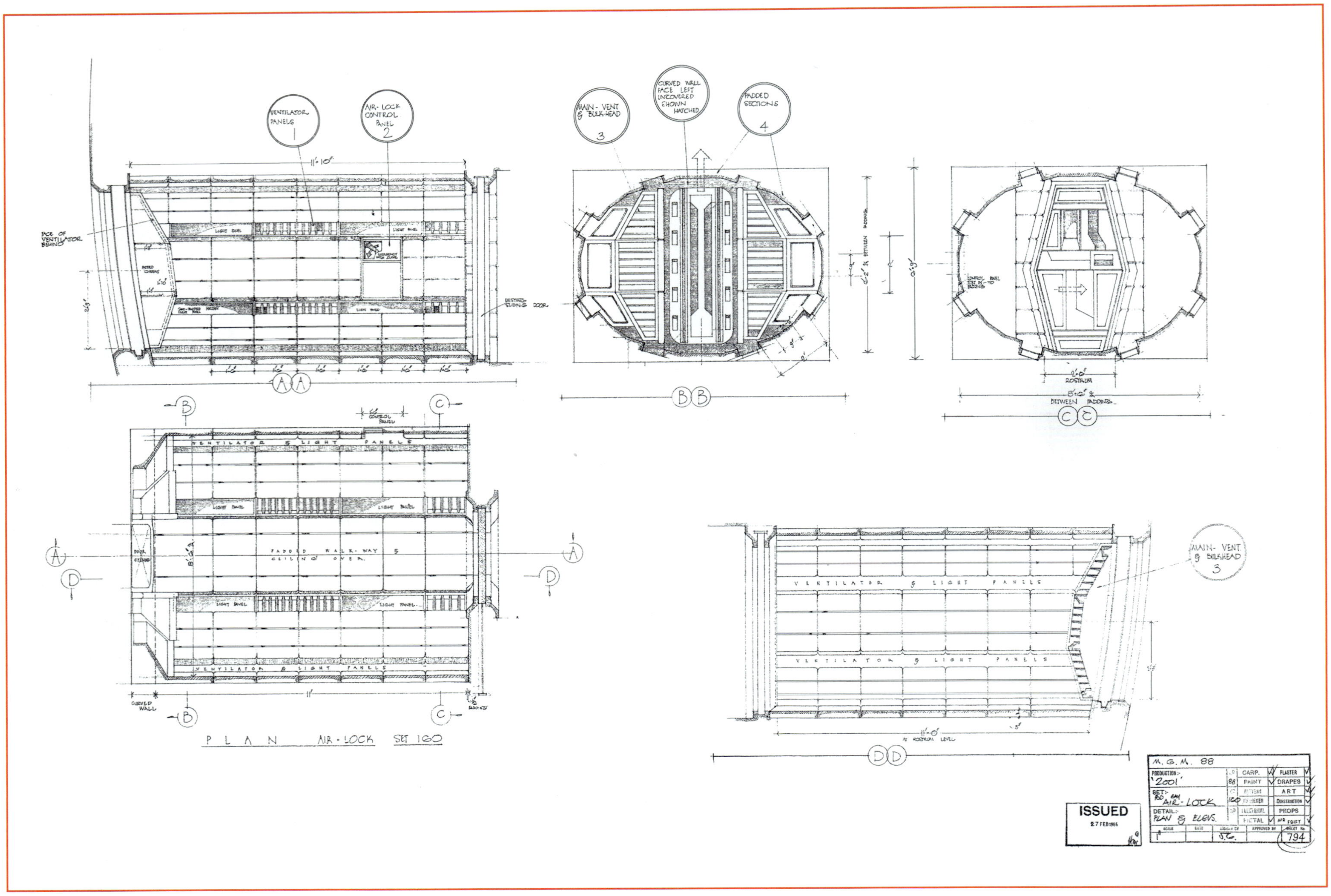

Air vents between Emergency lights can pressurize/depressurize area within 5 seconds. Control is completely manual and independent of onboard computer.

View looking out into space – view BB.
Air-lock control panel was relocated closer to exit, and an emergency EVA suit is stowed.

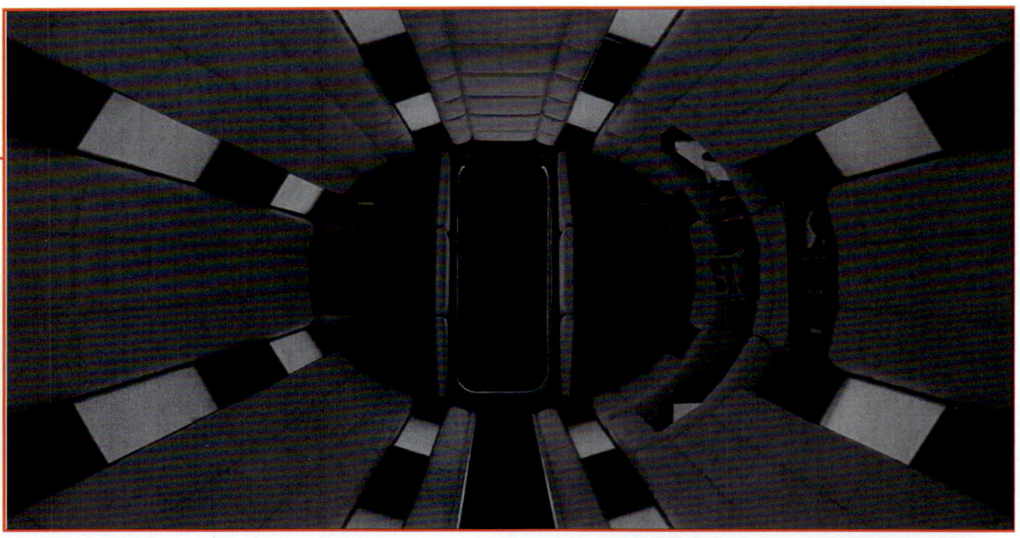

View looking into Podbay – view CC

The HAL9000 Computer

The HAL9000 (or *Hal*) computer runs everything on Discovery. Hal is an Artificial Intelligence computer with an optical memory core and heuristic algorithm design. Hal's functions on Discovery demonstrate he has to do about two trillion computations each second, and possesses a memory core of over a quintillion megabytes.

Artificial intelligence is a form of machine perception whereby computing machines can sense and interpret images, sounds, and contents of their environments, or the contents of stored media.

Famous twentieth century mathematician Dr. Irving John Good was consulted. Good's authorship of treatises such as "Speculations Concerning the First Ultraintelligent Machine" and "Logic of Man and Machine" (both 1965) made him the obvious person for Stanley Kubrick. In a 40 page interview with Good (probably conducted by Ordway), the HAL 9000 supercomputer was brought to life.

"Let an ultra-intelligent machine be defined as a machine that can far surpass all the intellectual activities of any man however clever. Since the design of machines is one of these intellectual activities, an ultra-intelligent machine could design even better machines; there would then unquestionably be an 'intelligence explosion,' and the intelligence of man would be left far behind. Thus the first ultra-intelligent machine is the last invention that man need ever make."

(From "Speculations Concerning the First Ultra-intelligent Machine")

Fred Ordway and Dr Irving John Good in early 1966

Hal's real-time perception of his environment is useful in industrial processes, such as assembly, inspection, diagnosis, vehicle guidance, engine status, oxygen levels etc. Off-line perception of stored memory is useful in medical, photo interpretation, content-based indexing and retrieval of stored experiences and media.

FUNDAMENTAL SPACE FLIGHT QUESTIONS	MISSION INFORMATION REQUIREMENTS	SPACE FLIGHT INFORMATION CATEGORIES	ATTITUDE	FLIGHT PATH
1. WHAT AM I DOING? (Orientation Information)	Definition of: > Present Position > Present Flight Path > Predicted End Conditions Orientation with respect to the present Flight Path.	> Guidance and Navigation > Flight Path > Time > Attitude	VERTICAL DISPLAY: > Present Attitude > Attitude Reference > Atttude Change	> Present Flight path > Predicted flight path
2. WHAT SHOULD I BE DOING? (Director Information)	Command: > Flight Path > End Conditions > Orientation > Present Position	> Guidance and Navigation > Velocity > Time > On-board systems > Attitude	VERTICAL DISPLAY: > Command Attitude > Command Rate of Change of attitude	> Command flight path
3. HOW AM I DOING? (Quantitative Information Expresses Magnitude)	Command: > Flight Path > End Conditions > Orientation > Present Position		VERTICAL DISPLAY: > Error between present and command attitude > Error between present and command rate of change of attitude	> Error between present and commanded flight path. > Error between present and predicted flight path. > Error between present flight path and flight path limits.
4. HOW SHOULD I BE DOING? (Quantitative Director Information)			VERTICAL DISPLAY: > Allowable tolerance in attitude and attitude rate	> Flight path limits

The IBM consultation group (Bevilacqua, Fox, D'Arcona) provided this basic programming structure for HAL. The perception of HAL being 'sentient' is provided by four fundamental philosophical questions, and their relevance to the mission.

5. INFORMATION AREA	TASKS
A. ATTITUDE:	1. Obtain attitude requirements for change in trajectory 2. Control attitude as required 3. Monitor attitude with respect to inertial space (from inertial platform) 4. Fine alignment of inertial platform
B. PROPULSION	1. Manage thrust profile 2. Monitor thrust shut-off 3. Compare required propellant for orbit change with amount of remaining mission 4. Countdown to t-0 for thrust vector management 5. Initiate ignition sequence 6. Monitor for normal ignition 7. Monitor for normal operation

#6 is an example of a specific function the crew of *Discovery* would ask and expect detailed follow-up.

6. SUB FUNCTION: TELESCOPE	PERFORMANCE REQUIREMENTS
Determine overall system effectiveness in terms of mission goals.	high-order decision based on trade-off of system status, component by component.
a. Convert accelerometer and data to digital input. b. Input into computer as delta vector velocity, maneuver takes place.	>Conversion should have minimal error, (intrument errors are on the order of 0.02%/hr.) >Real time input and address selector.
a. Select address, process routine (instruction stored there). b. Oversee routine (display of updates trajectory determinants, and acceleration vector).	> Enter address number or designator in input of control element. > Compare with **ground report**, dead reckoning and independent range estimates.
a. Select addresses for the acceleration vector b. Storage	> **Pre-Arranged** before mission. Implicit to routine.
a. Operate scan telescope b. Locate approximate area of celestial object of interest.	> **Ability to discriminate** star patterns and/or other celestial objects.
a. Gross control of attitude to bring sextant facility within viewing range. b. Slew sextant to bring object within field; repeat for second. c. Stabilize vehicle to keep object within field during routine (function like tracking with 200" Palomar telescope).	> Gross nulling task, not usually time crtitical. > **Ability to discriminate** star patterns and/or other celestial objects. > Minimum rates during period of observation approximately 1sec/sec. A tracking-compensatory-task.

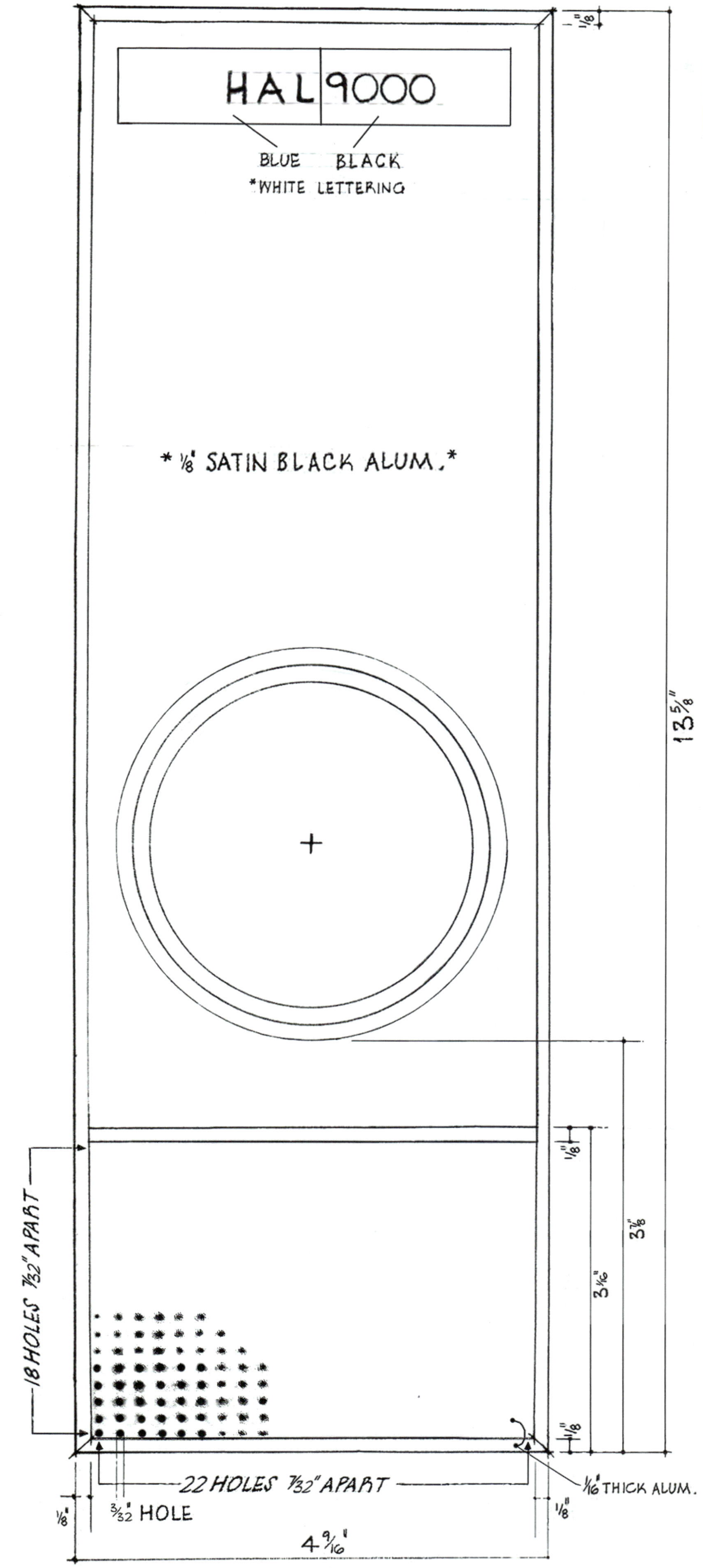

HAL 9000

These measurements were taken directly from the 'ring' artefact. The hole is cut-out for a Nikon 8mm F/8 lens.

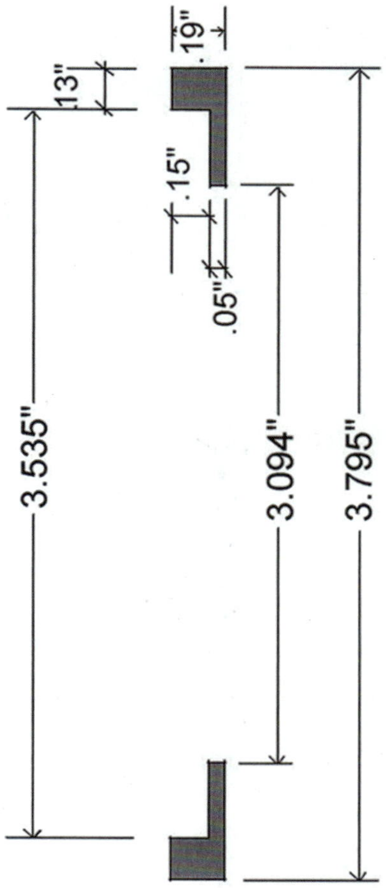

- BLUE / BLACK
- *WHITE LETTERING
- * ⅛" SATIN BLACK ALUM. *
- 13 ⅝"
- 4 9/16"
- ⅛"
- 3 ⅜"
- 3 3/16"
- ⅛"
- 1/16" THICK ALUM.
- ⅛"
- 3/32" HOLE
- 18 HOLES 7/32" APART
- 22 HOLES 7/32" APART

Ring measurements:
- .19"
- .13"
- .15"
- .05"
- 3.535"
- 3.094"
- 3.795"

Early AI researchers developed algorithms that imitated the process of conscious, step-by-step reasoning that human beings use when they solve puzzles, play board games, or make logical deductions. By the late 1980s and 1990s, AI research had also developed highly successful methods for dealing with uncertain or incomplete information, employing concepts from probability and future certainty equations based on current real-time events.

For difficult or contradictory problems, most of these algorithms can require enormous computational resources — most AI experience a "combinatorial explosion": the amount of memory or computer time required becomes astronomical when the problem goes beyond a certain size. By design, a heuristic algorithm is an algorithm that gives up finding the optimal solution to improve run time. A heuristic abandons one or both of these goals. *Note: In the film, Hal experiences paranoia and killed the crew of Discovery in an attempt to contain the true nature of the mission. This is an example of a combinatorial explosion.* Major AI textbooks define artificial intelligence as an intelligent agent where a system perceives its environment and takes actions which maximize its chances of success. AI can be seen as the realization of an abstract intelligent agent (AIA) which exhibits the functional essence of intelligence.

Among the traits Hal exhibits are reasoning, knowledge, planning, learning, communication, perception and the ability to move and manipulate objects (such as the Pod and its manipulator arms).

Hal's researchers used tools and insights from many fields, including computer science, psychology, philosophy, neuroscience, cognitive science, linguistics, ontology, control theory, probability, optimization and logic. AI research also overlaps with tasks such as robotics, control systems, scheduling, data mining, logistics, speech recognition, facial recognition and many others.

The 'brain room' is located on the top level of Discovery, starboard side. Dimensions (internal) are 15' x 15'6" x 3'10".

The "brainroom" located between the command deck and the centrifuge, uses banks of an optical memory element capable of recording huge amounts of data provided by a clear thermoplastic (heat-deformable) resin memory sheet between an array of nano-lenses. One of these lenses, for example, focuses readout optical radiation incidents on a single desired page of the memory sheet. The readout optical radiation comes from one of millions of light sources, each of which correspond to a page of the memory sheet. The other nano-lens focuses a page of write-in optical radiation onto the corresponding page of the memory element. This second lens also focuses the readout optical radiation after propagating through both the memory elements. An array of optical detectors, onto which the readout beam of optical radiation is focused, reads out a page of information at a time, corresponding to the particular readout optical source energized at that time.

The HAL 9000 optical memory storage units.

HAL continuously monitors the following systems: These are displayed to the personnel on 8" x 8" monitors that are in almost every room of the habitat sphere. The 3-letter acronyms for HAL were created by Douglas Trumbull.

ATM: Atmosphere check. Oxygen levels and Air pressure in the habitat areas.

COM: Communications status check via various outboard antennas.

CNT: Coordinated Network Test. Check all connections of the HAL network throughout the ship.

DMC: Data Management Control check.

GDE: Geospatial Data Exploitation. Interpretive analysis of planets and other space objects Using topology and cartography derived from surface scans.

FLX: Flex Physical Simulation. 3D interactive application that converts the GDE mapping data.

HIB: Hibernation chambers status check.

MEM: Optical memory storage check.

NAV: Navigation status check.

NUC: Real-time monitoring of the nuclear plasma core.

VEH: Status of the three Pods.

MRN80-EJ ATM

FMI40-Q1 COM

REA6W-T5 NUC

This display shows course plotting over time and distance between Earth and Discovery's destination.

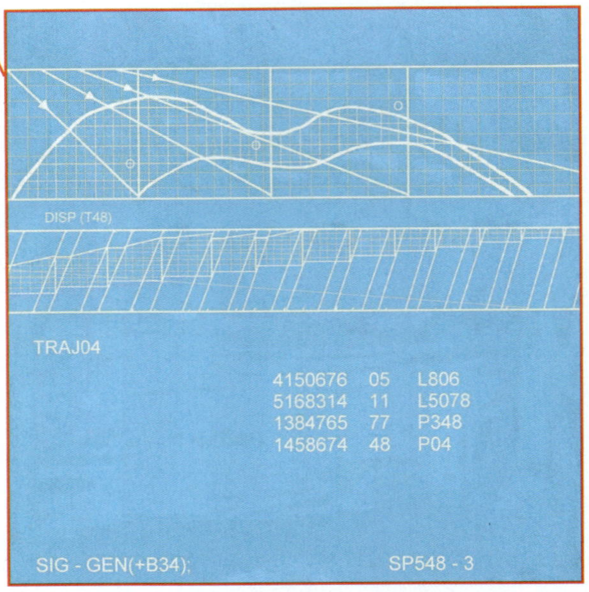

This display shows a fundamental 3-D course plot check.

This display is part of the DMC overall ship systems management.

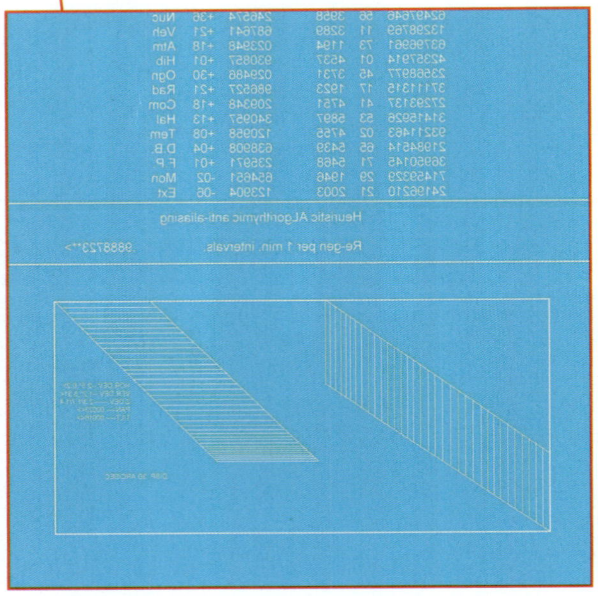

This display demonstrates HAL's ability to create any architectural part of the ship in 3-D and precisely locate and test critical system anomalies.

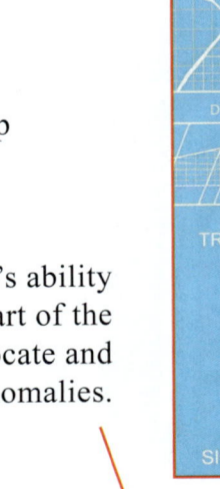

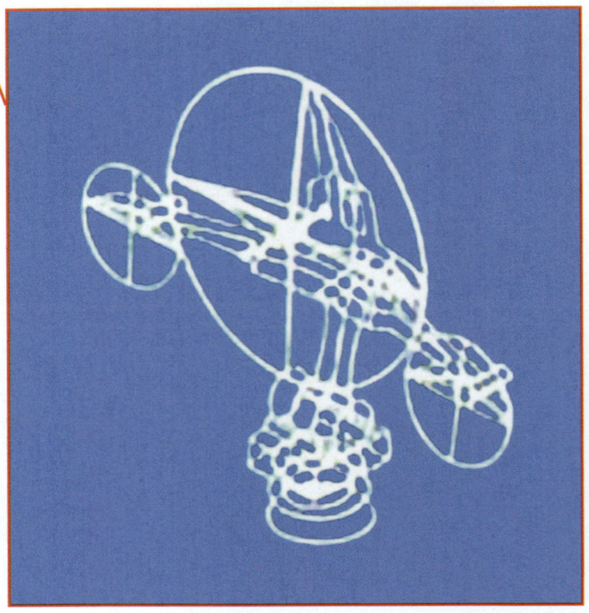

THE BEST MODEL SHOP IN FILM HISTORY

Craftsmen hired for the miniatures department were of exceptional skills in their given areas of expertise. Only those with a superb portfolio and the ability to follow a drawing exactly were hired. Talents included millers, wood-lathe operators, metal-lathe operators, and architectural modelers. Models for the film were created directly from 1:1 scaled drawings of all the vehicles. The only exceptions to this were: large *Discovery* at 3" to the foot, and *Aries 1B* at 3X drawing size. This way the model shop craftsmen could put their calipers on the drawing and direct reference it to the part being worked on. Andrew Birkin writes: "I was really just the tea-boy-who-got-lucky, being 19 when I started work on the film at £8 a week. Having caught the Great Man's ear by pretending that I knew of a desert in England (I didn't, but managed to rustle one up pretty fast!), he elevated me to the lofty-sounding position of First Assistant on Special Effects. My basic task was to co-ordinate the five special effects crews that were working 24/7, as well as organize the modelers, and every few days I would give Stanley Kubrick a progress report of the models along with Polaroids".

The Titov V model in the final phases of construction. This model was in scale to the Orion III at about 48" in length. The model was hollow fiberglass construction and painted anti-flash white (a paint used by aerospace manufacturers on jets in the 1960's to protect the crew from the initial flash of a nuclear bomb detonation). Many gallons of this paint were purchased to cover the enormous model department's giant-scale work. Vickers Ltd. helped in supplying the paint since MGM was already contracted with them.

July 25 1966

August 10 1966

First version of Orion III had a large engine 'hump' in the rear and large cockpit window inclined at 70 degrees. Kubrick and Masters had the 'hump' reduced in size and the cockpit window angled to ~ 50 degrees to enhance the aerodynamic accuracy and maintain scientific realism.

Underside of 42" Orion III model is having final coats of panel shading. The model was hollow fiberglass with acrylic wings.

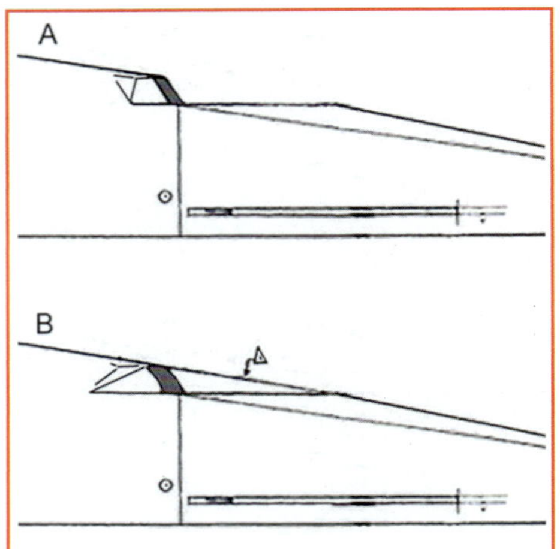

July 25 1966

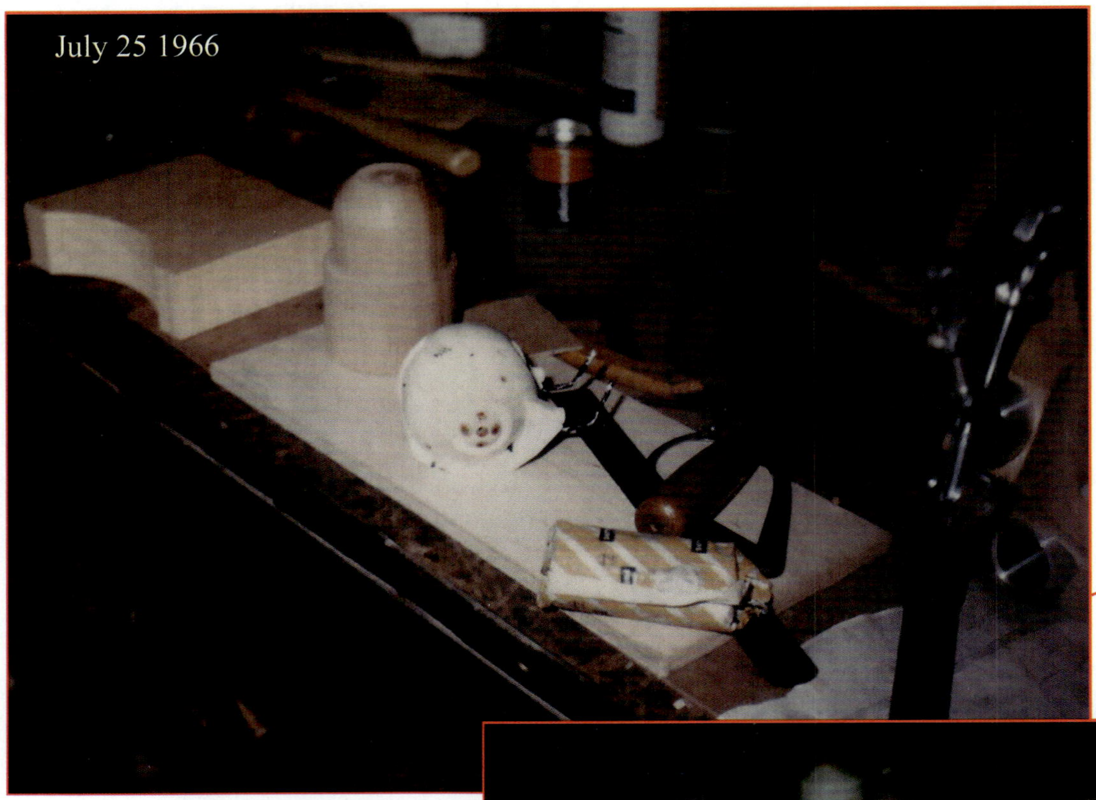

Smallest pod model was 3" in diameter. Built of a machine lathed acrylic hollow ball and hand-crafted details.

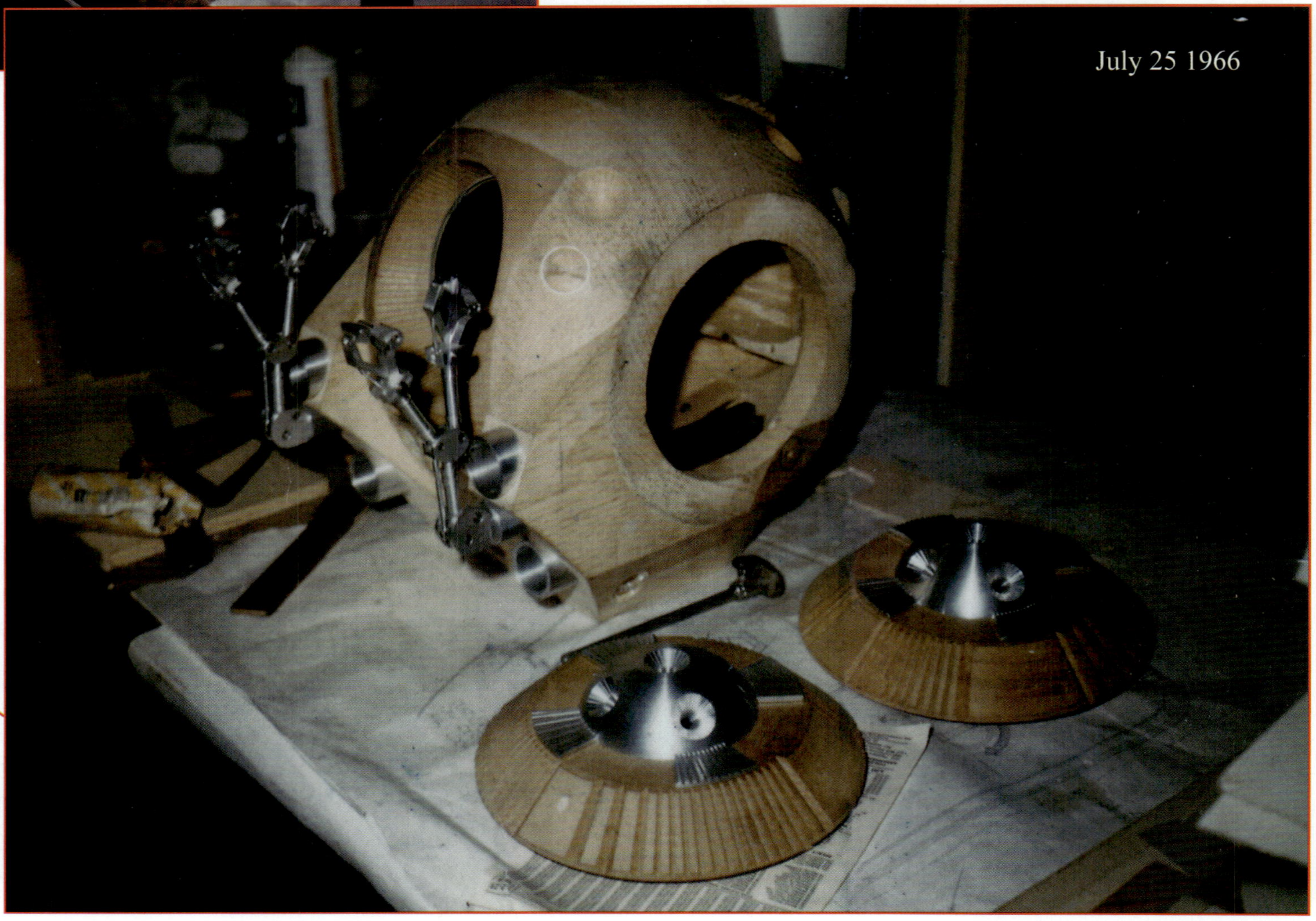

A. First version of window titled @ 70°

B. Final version of window tilted @ 50° and, slightly curved. Note area behind window cut deeper in fuselage.
The 'Delta' symbol indicates to model shop that center window must conform to body contour.
The left and right window corners stick out beyond the contour of the body.

Four pods were ordered from the model shop, and only two sizes were actually created due to time constraints and the time involved creating only one. This one is 13" in diameter, and built of many pieces of hardwood glued together. Precision machined aluminum detail and arm/hand controllers were milled.

July 25 1966

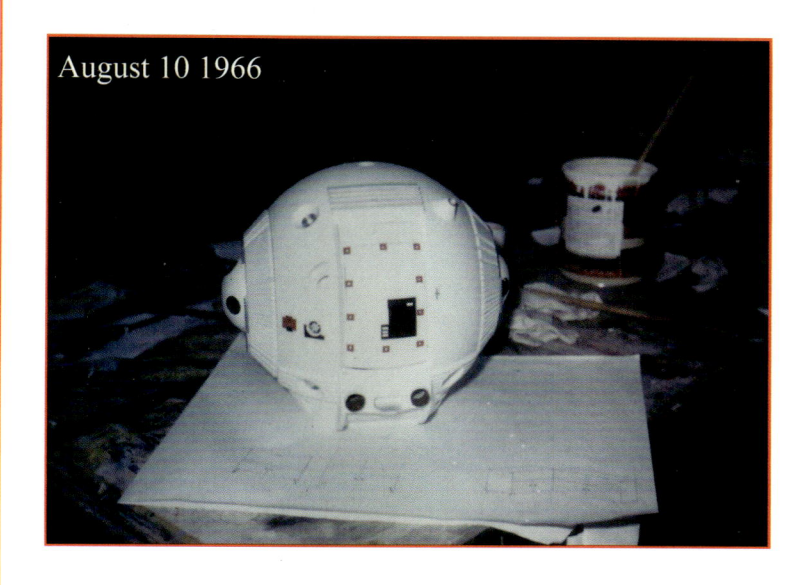

August 10 1966

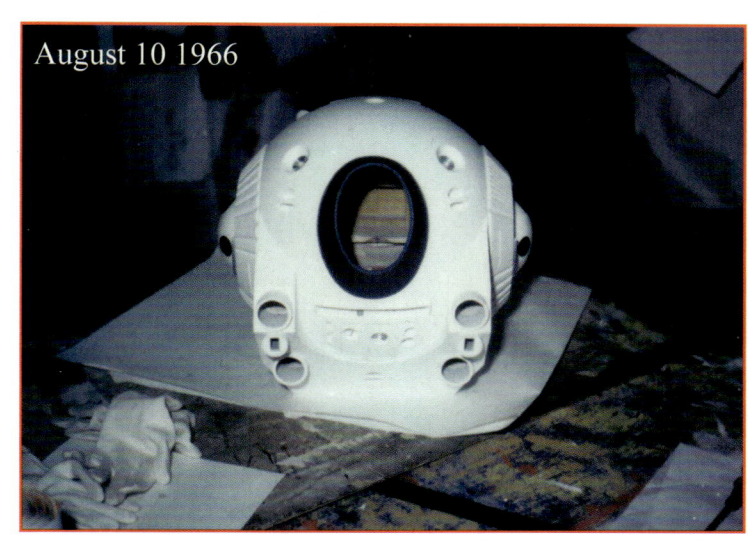

August 10 1966

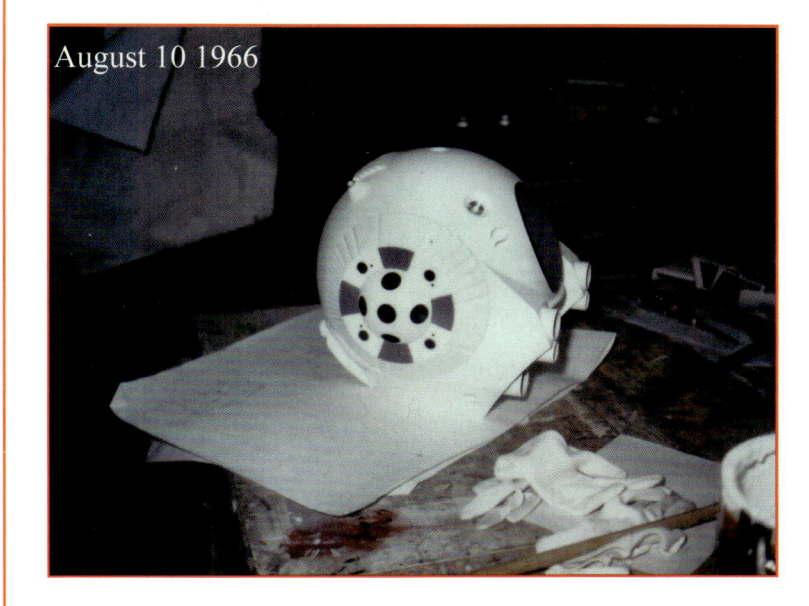

August 10 1966

July 25 1966

July 25 1966

August 31 1966

The pod sat on the egress arm that was built over a foot in front of the large 54' Discovery model. The pod was over-scaled by 3" and had to be photographed with forced perspective to make it look in-scale to the discovery model. Other shots of the pod were photographed separately and super-imposed in the film to be visually in scale. This oversight was due to the Podbay doors on the 54' Discovery miniature being oversized by 3".

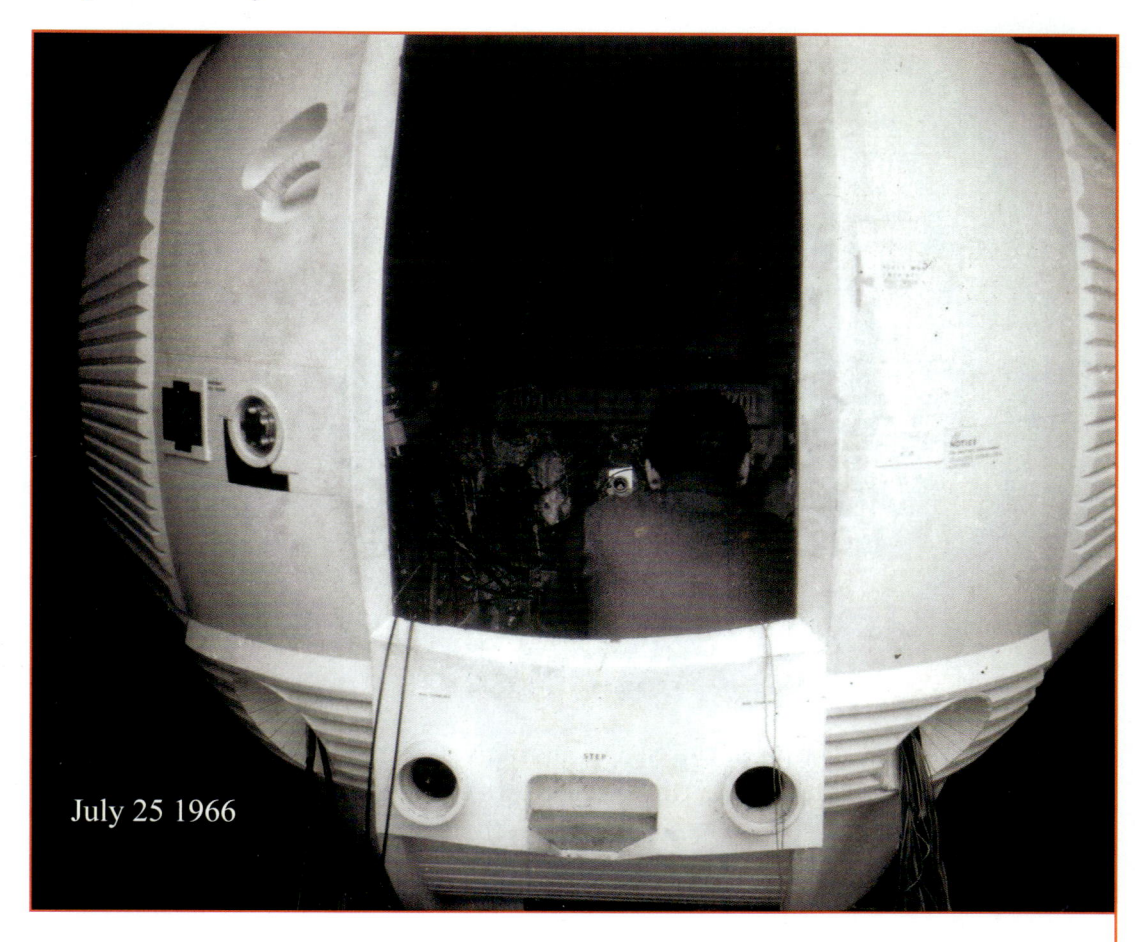

A technician is in the full-scale pod that was wired specifically for control arm actuating.

The aluminum hub of the Space Station is turned on a lathe - a hollow construction makes it lighter and able to be handled. This part was about 15" by 39". The two cylinders to the right are inserted into each end, for mounting to a motorized armature that bolts to either end of the hub for filming purposes.

A view of the complex of electric motors that actuated the fine 'arm' and 'hand' movements.

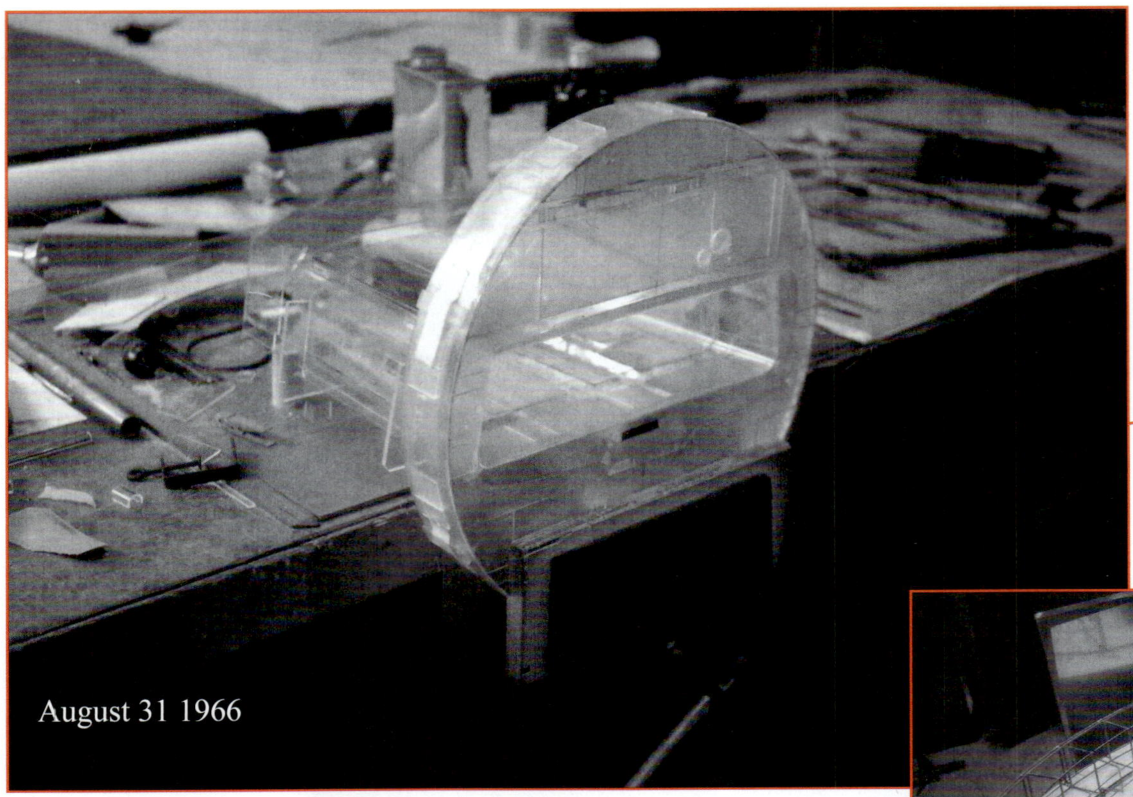

The entry port for space shuttles is built entirely of precision milled and cut pieces of acrylic. This part 'snapped' into place on either end of the model.

August 31 1966

August 9 1966

The 'incompleted' section of the Space Station was constructed of many die-cut pieces of copper spot-welded together. Steel piping comprises the structural core of the giant 8 foot wide model.

The 8 foot diameter main rim of the Space Station is 75% completed. Many sections bonded together are made of fiberglass.

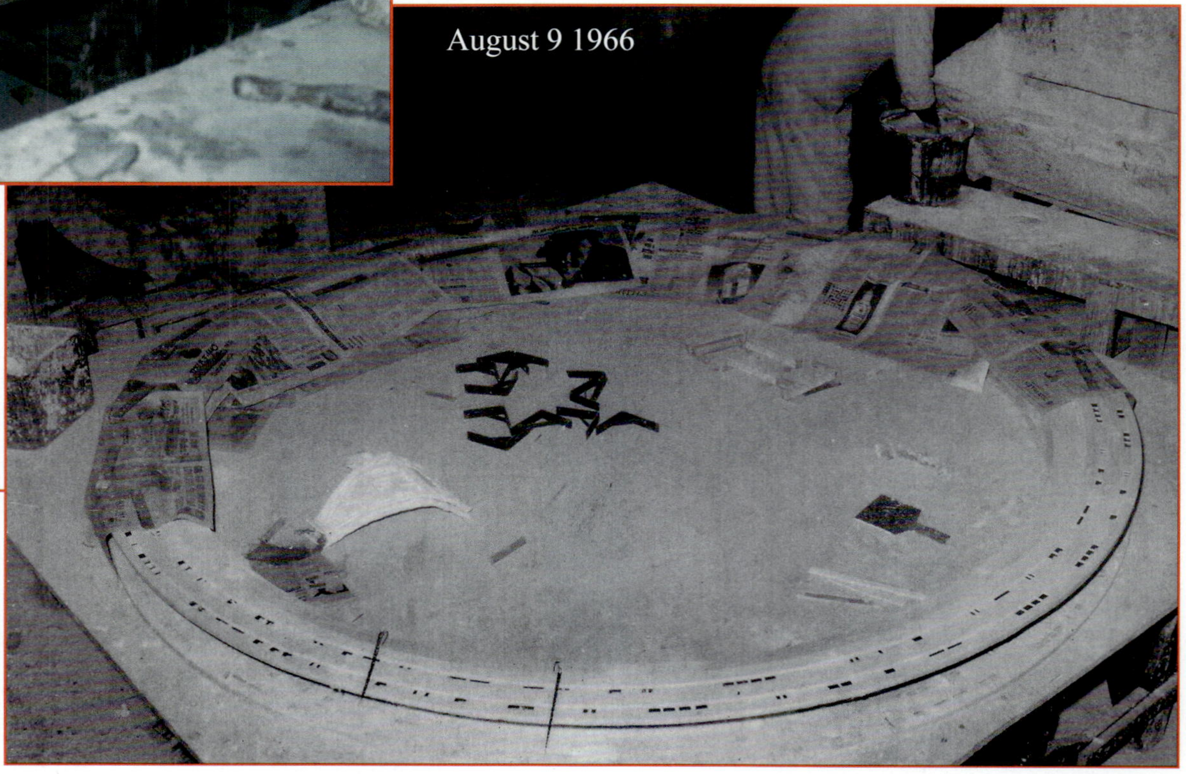

July 25 1966

August 9 1966

A lighting grid was placed inside before outside fiberglass plates were secured.

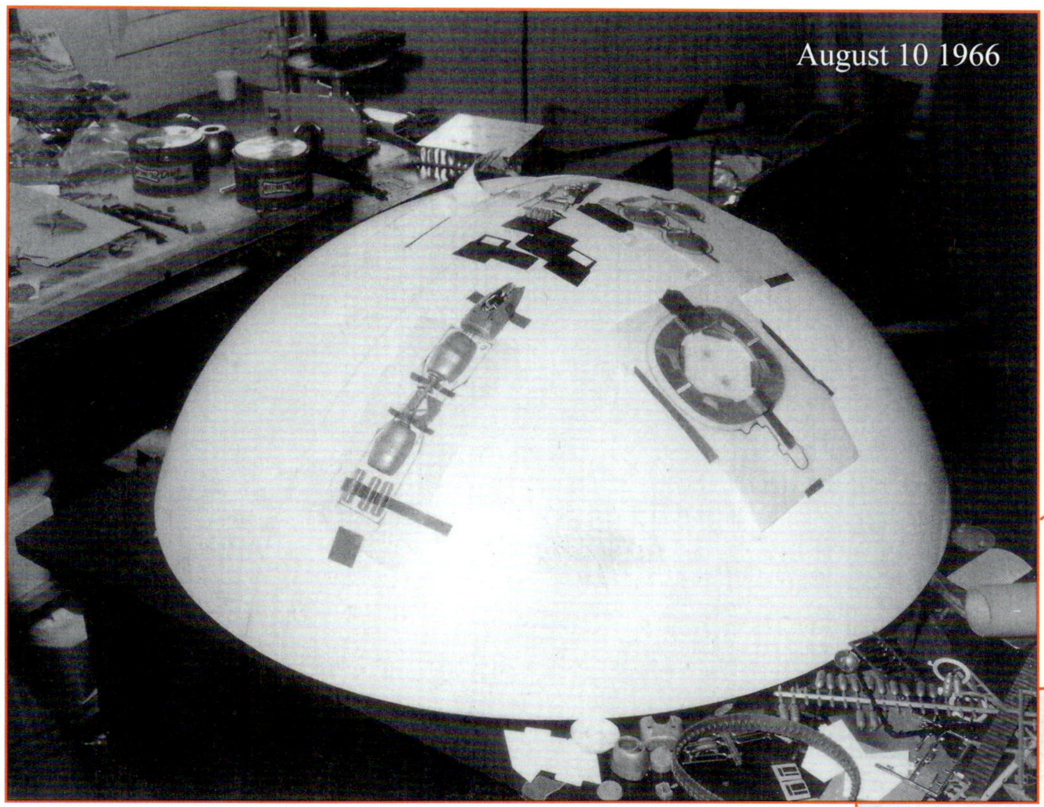

August 10 1966

The top dome of the Aries 1B has many pieces of model kits taped into position for ideas of surface detail. This mock-up was rejected by Kubrick and Masters in favour of raised panels.

Landing gear legs were redesigned many times over. The left side shows a rejected design, the right the final design. The final design was never drawn – only fabricated.

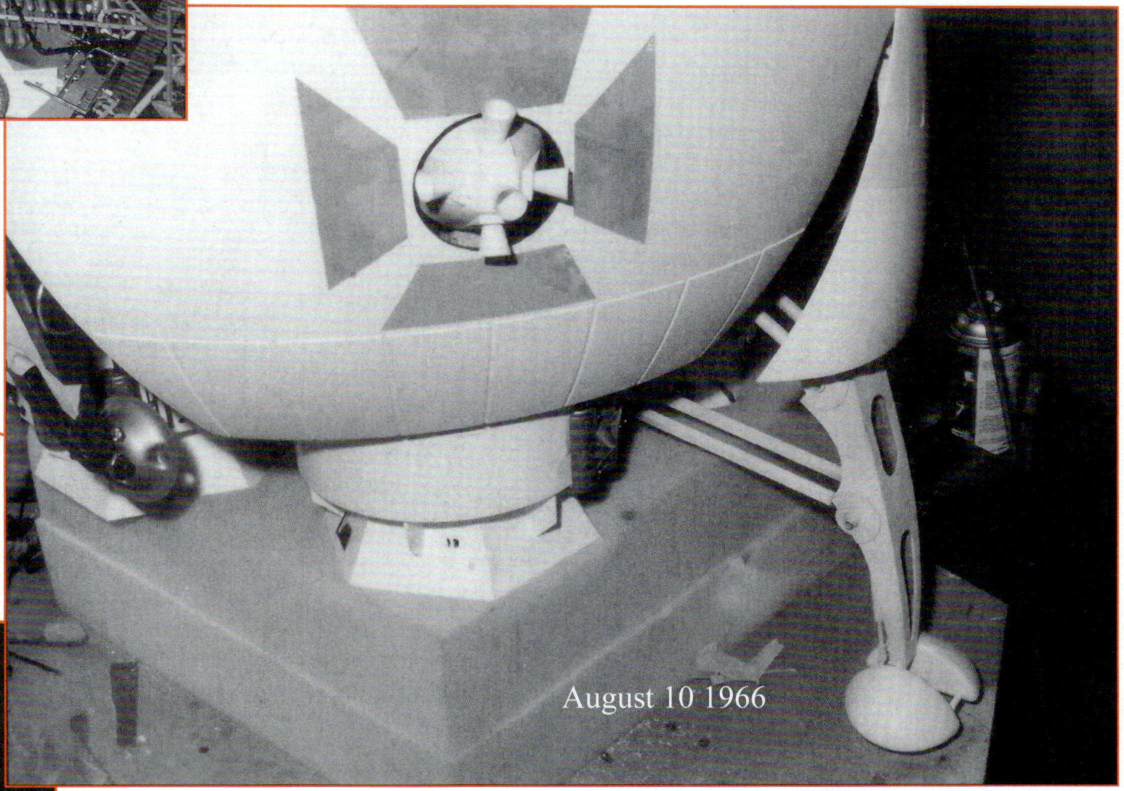

August 10 1966

July 25 1966

The Aries 1B was a 3' wide fiberglass sphere. Here, engine details are fashioned from many model kits and copper wire/tubing.

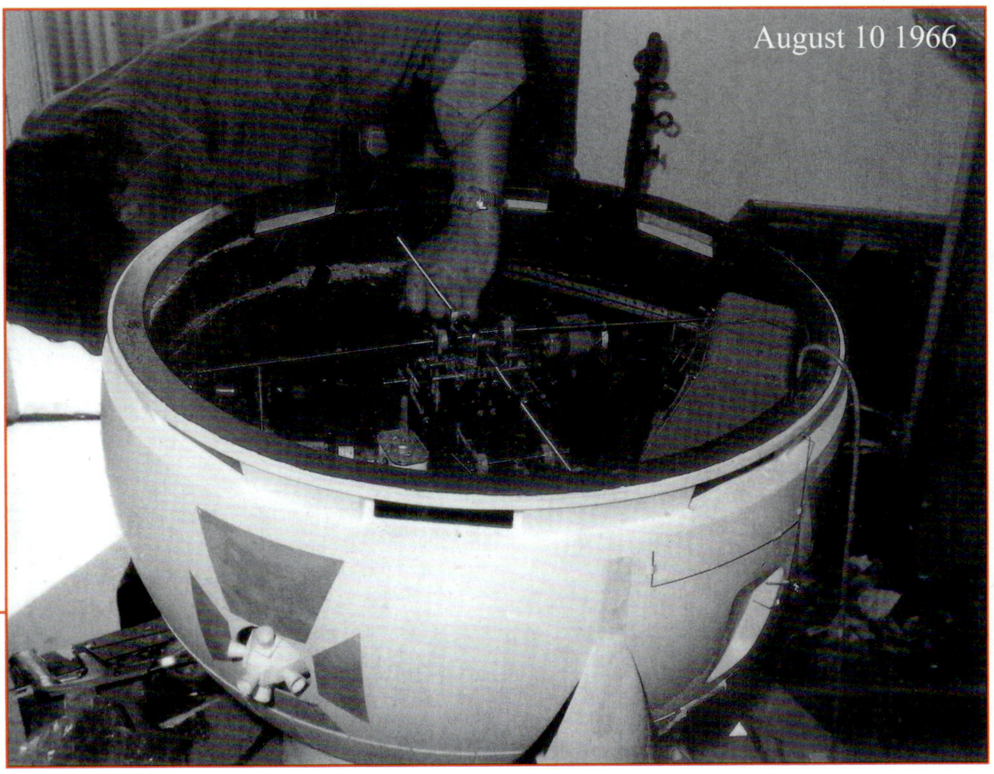

August 10 1966

The inside of the Aries 1B shows the four-way motor actuator system that lowers the landing gear system. An electric regulator could program the start and stop position of the legs.

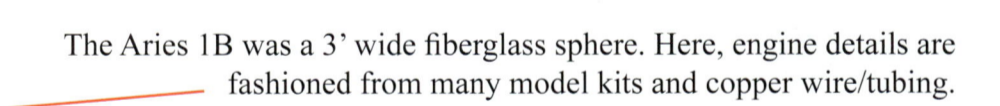

91

August 9 1966

August 31 1966

August 9 1966

The smaller Discovery model was 15' long. The neck piece (behind the sphere) is turned on a wood-lathe, and is fashioned out of several pieces of hardwood clamped and glued together. (above)

The Reactor and pressure plate sections are complete, other than surface details which again, are model kit parts. The 'collar' piece sits in front of it. Internal frame and bracing are hardwood, and the flat sections are plywood. (left)

The spine and complete nuclear engine assembly are ready for detailing. Elliptical piece at the rear was rejected quickly. Its intention was to protect the nuclear housing from the initial blast wave after primary ignition from earth orbit. (top right)

The Command sphere is two halves of an acrylic ball. All the details are precision cut and milled pieces of acrylic. (right)

Tankage sections are assembled, and one of the aluminum connecting rings sits on a table. (below)

After a coat of primer. (below right)

August 9 1966

August 9 1966

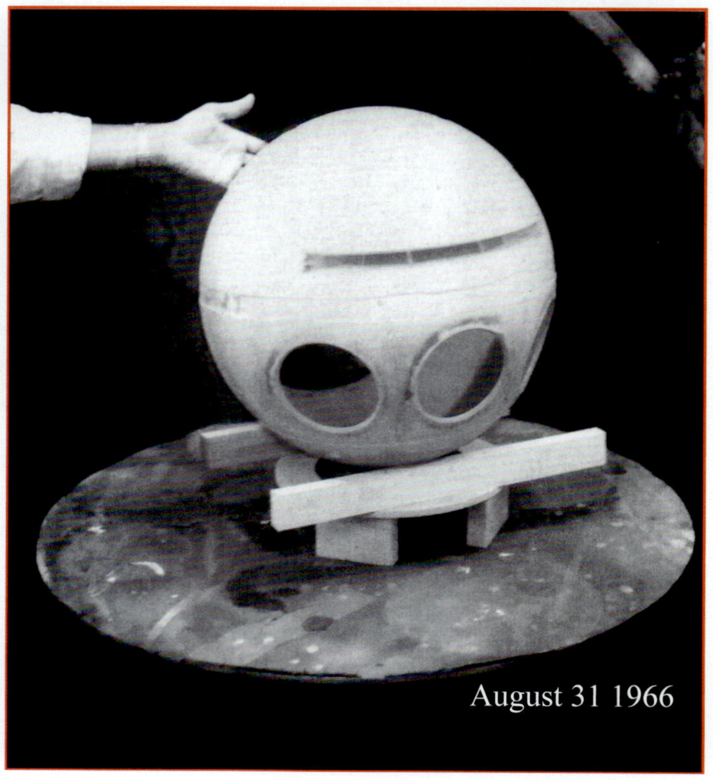

August 31 1966

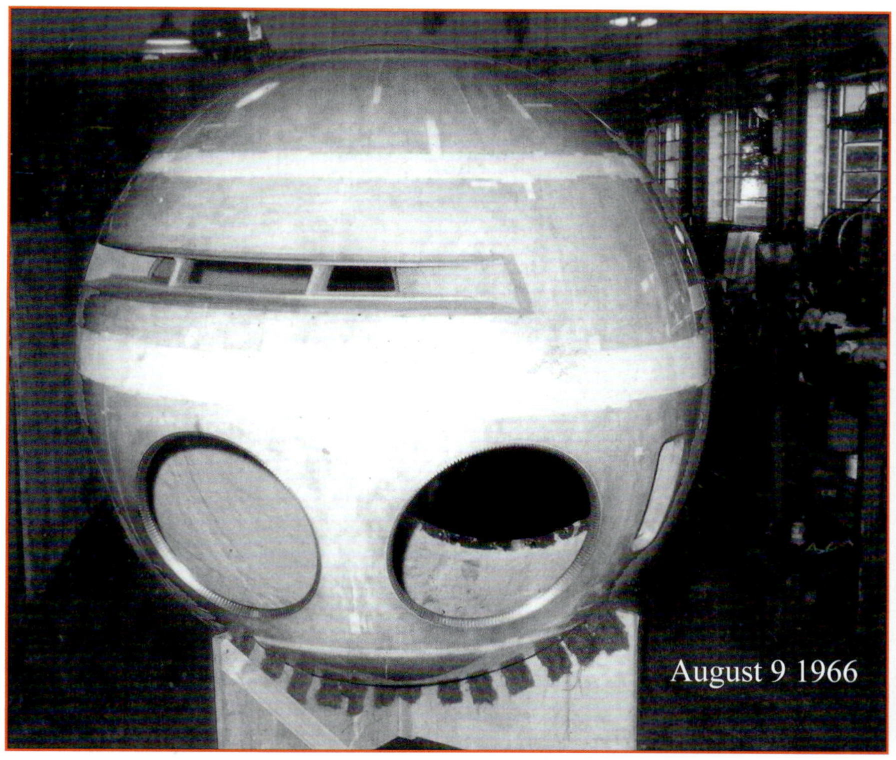

August 9 1966

The large Discovery model was 54' long. Still the most expensive and largest miniature ever constructed for a film. The pressure sphere was 6' 0" in diameter and was constructed of dozens of pieces of milled hardwood. Surface details were a specialized thin metal that could conform to contoured surfaces.

July 25 1966

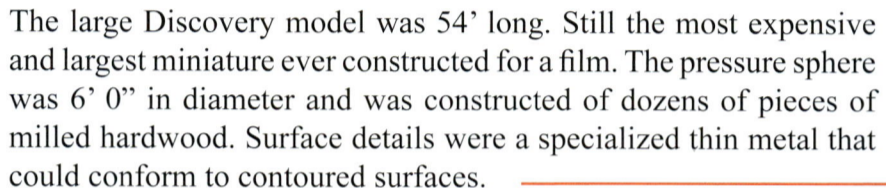

The drawings had the podbay doors sized at 13' 0" (as viewed from the outside) relative to the miniature being in-scale with the interior drawings. The interior drawings show the podbay doors as 10' 0" (as viewed from the inside). This may have been a simple oversight by the engineering team who interacted with the draughtsmen.

August 9 1966

July 25 1966

A wood craftsman sands the end pieces of the plywood tankage sections.

The neck and collar of the Discovery was about 4' by 4'. Again, fashioned from many pieces of hardwood. Surface details (not yet in place) were of milled acrylic, and small model kit parts.

Tankage sections lay assembled and ready for mounting to a metal tube core.

August 9 1966

A steel tube is the core of the Discovery. Here, one tankage section plus two of the rings have been placed.

August 9 1966

August 31 1966

A completed coupler unit sits on the table for review. Right is the ring of the coupler. The ring is acrylic, the coupler is machine lathed aluminum.

August 9 1966

Rings were removed after entire 54' model was completed when Kubrick decided he didn't like the look of them.

94

August 9 1966

Huge models of the lunar surface up to 30' X 30' were also made and carefully matched to photos. Visual effects supervisor Douglas Trumbull was one of the primary sculptors of these lunar landscapes. The Soviet Embassy in London provided photos of the lunar surface from the Zond and Luna probes that actually landed on the moon. Although poor in visual quality, they provided a good idea of the lunar terrain and mountains as viewed at ground level.

A giant 8 foot model of the moon was created and surface details sculpted from grey colored plaster.

August 10 1966

Fred Ordway holds pictures he acquired of the moon taken at the Pic du Midi observatory in France. In 1966, these were the best pictures of the moon available.

APPENDIX I:

Production photos and notes.

Stanley Kubrick and Directory of Photography Geoffrey Unsworth developed a system of photographing the miniatures to look as real as possible. These photos were originated with an 8" X 10" plate camera.

This series of lighting tests was conducted with the miniature sitting in front of non-reflective black velvet material. Absolute control of front and fill light could be achieved with this method. It was discovered during this phase that lighting the object from below or behind gave a sense of great scale; as compared to front lighting which made the miniatures look dull and toyish.

In this example of four independent photos of the *Orion III*, Stanley Kubrick picked the uppermost photo for use in the film. It was then re-photographed in color with the same lighting conditions and implanted into the final 70mm film print.

F-22 - No Backfill

F-16 - 20% Backfill

F-12 - 20% Backfill

F-8 - 30% Backfill

Fred Ordway and Harry Lange in Stanley Kubrick's office at Borehamwood Studios discussing the day's schedule.

Stanley Kubrick, Fred Ordway and model Maggie London in the Space Station V set

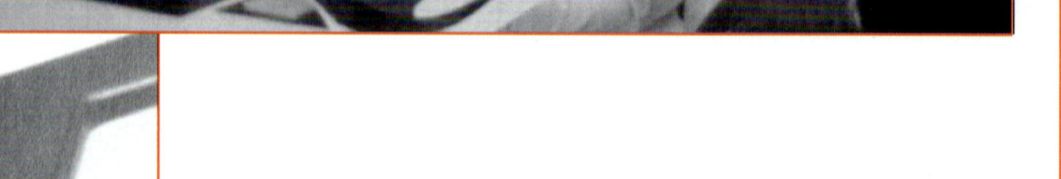

Fred Ordway at the Reactor Controls in the centrifuge

Fred Ordway and various production staff in the Space Station V set

Construction manager Dick Frith, Harry Lange and Production designer Ernie Archer at 'Master Models', a company in England that produced models and sets for the film. They are working on insert storage receptacles that were in the storage room. Never seen in the final film.

Plate camera lighting test of the Rocket Bus. Note that the windows have black velvet tucked in them.

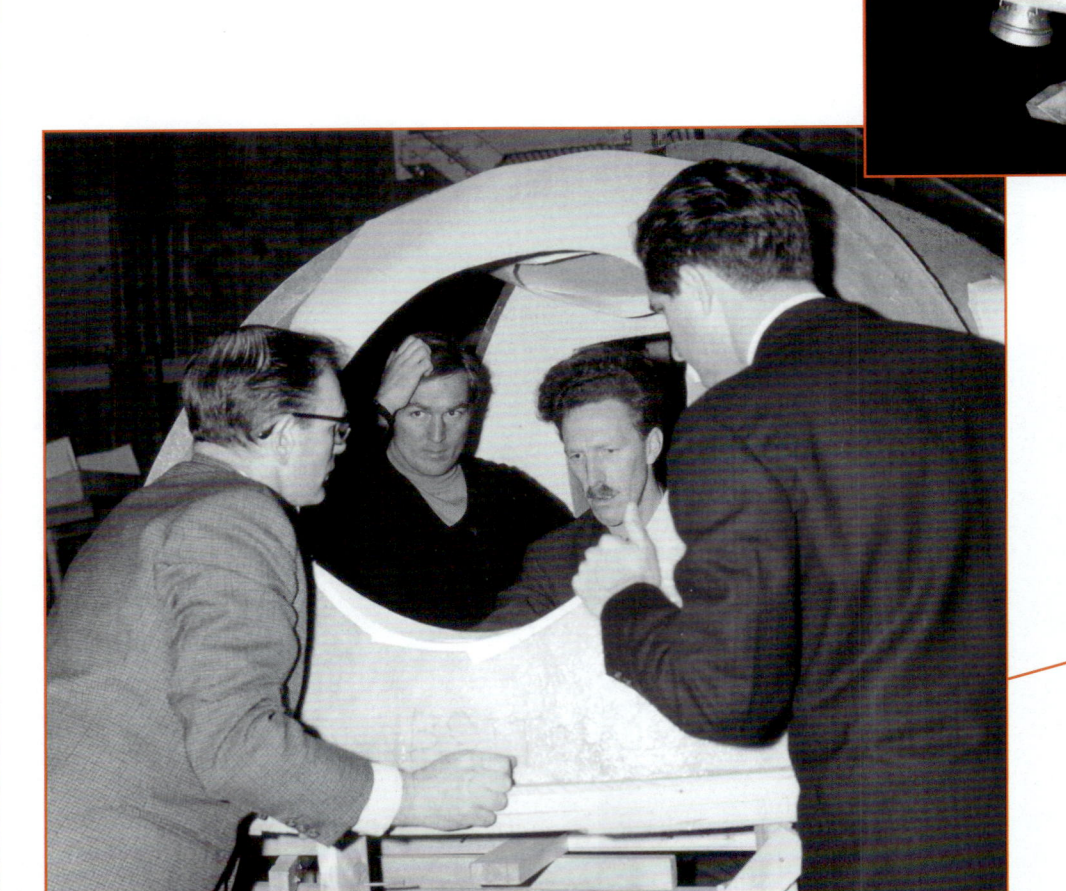

Fred Ordway with the Hawker Siddeley Dynamics consultants trying to figure out how to pack all of the instrumentation into the EVA Pod (half completed).

Ordway stands in the corridor of the space station where automated machines dispense gifts, drugs and cameras. Production designer Masters wanted the Space Station to have the look and functionality of modern airports.

Ordway and Lange discuss details of designs during construction of the centrifuge

Ordway directs model Maggie London on how to use the console in the space station import control area.

Ordway, Masters and Lange discuss design and engineering details of the many production pieces.

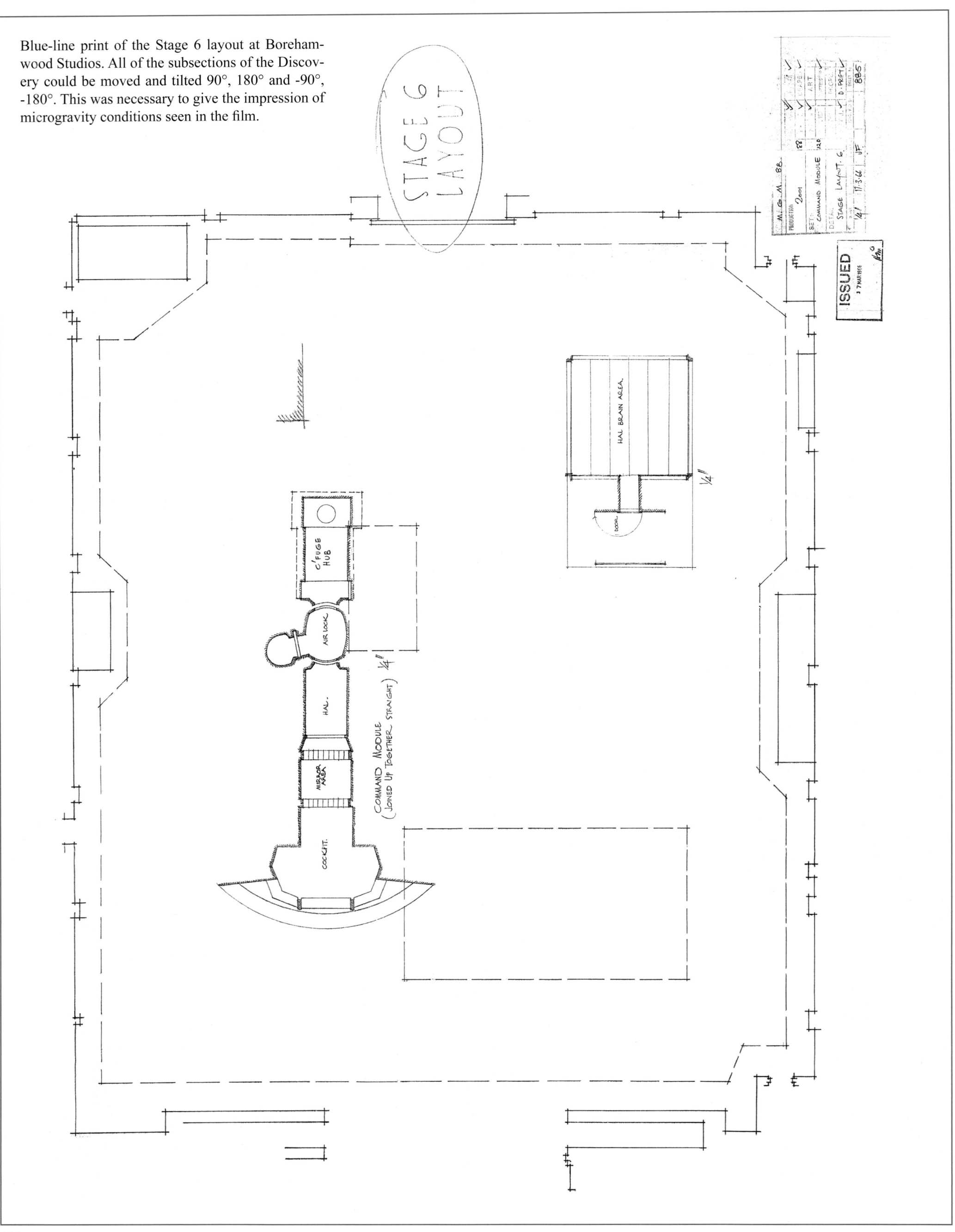

Blue-line print of the Stage 6 layout at Borehamwood Studios. All of the subsections of the Discovery could be moved and tilted 90°, 180° and -90°, -180°. This was necessary to give the impression of microgravity conditions seen in the film.

"Backstage Magic For A Trip to Saturn" - April 1967

Earliest known publicity was this writing by contract author Richard F. Dempewolff. Arthur C. Clarke gave him a rare tour of the MGM British studios in early 1967. Dempewolff interviewed Clarke and Ordway. Although some details changed after the interviews later in production, the core had some unique insight.

ARTHUR C. CLARKE, noted authority on outer space, picked his way carefully across the surface of the moon a few paces ahead of me. Off to our left was the crater Tycho.

"You're standing on the landing pad for the Aries IB moon shuttle," Clarke called back over his shoulder. "Be careful not to step on any moon base buildings. They're plaster. They'll bust."

We were exploring a 30-by-30 model of the moon's landscape on top of a three-story-high scaffold in MGM's Borehamwood studios, near London. Far below, on the floor, several full-scale spaceship command modules were under construction, bristling with instruments, buttons and multicolored lights. For months, in the sprawling complex of studios, cameras had been filming eerily lighted sequences of spectacular vehicles moving through the solar system; interiors of orbiting spaceships in which "weightless" people walk up walls and cross ceilings upside down; fantastic scenes in a pin-wheeling 42-foot centrifuge inside Discovery - a deep space probe on its way to Saturn with a volunteer crew of one-way explorers.

I was witnessing the wrap-up of a gargantuan three-year effort by director Stanley Kubrick and an army of experts to produce the forthcoming adventure movie entitled 2001: A Space Odyssey. The film story, written by Kubrick in collaboration with Clarke, a famous science fiction writer, is based on a book Clarke wrote called The Sentinel.

Scientists determine that the direction of the monolith's signal indicates life in the Saturnian system. Earth people already have plans for a deep-space ship, Discovery - a giant affair 600 feet long powered by fission plasma drive. The crew rides in a 50-foot sphere at one end. A 42- foot centrifuge within the sphere provides gravity for the men.

Planned for nearer planets, Discovery has only a one-way capability for Saturn. But the decision is made to go anyway. Five volunteers are picked for the voyage. During the year-and-a-quarter trip, traveling five million miles a day, they rotate duty - two on and three off. Those off duty hibernate in cryogenic (deep freeze) "coffins."

What the explorers find is enough to make flesh creep-and so were the many problems of staging this far-out extravaganza. Kubrick and Clarke insisted that everything in the picture had to be based on known, workable principles. Controls had to be realistically accurate, and many things actually had to work.

Since no one has yet been to the moon, much less to Saturn, it wasn't easy. The team that bore the "feasibility" burden included three indefatigable wizards: Fred Ordway, scientific consultant; Harry Lange, the designer who took Ordway's technical data and drew plans for workable space gadgetry no one had ever seen; and Tony Masters, an ingenious builder who turned plans into hardware.

So precise and logical are 2001's space vehicles that the National Air and Space Museum has requested them for permanent exhibit when the picture is finished. Detailed "specs" and drawings for an advanced fission plasma rocket capable of sending Discovery to Saturn were prepared by Thomas F. Widner, General Electric's manager of advanced nuclear programs. IBM development people, under Elliott Noyes, made workable plans for a fantastic red-lighted computer for the lunar module-exactly as it would have to be for a lunar computer. With Minneapolis Honeywell experts they designed a whole family of computer units adapted to spaceship use along with accurate panel displays and controls for every console shown in the film. From Bell Labs came authoritative advice on communications questions.

"The hibernation sequence was a beaut," Ordway recalls. Doctors at New York University Medical Center and Harbor General Hospital in California, familiar with work in the deep freezing of live animals, helped in the design of human cryogenic hibernators. In the movie, you'll see men climb into plexiglas containers where they are frozen to sleep.

Efforts to maintain scrupulous authenticity were endless. When detailed contour charts and maps of the moon were needed, Ordway couldn't find any good enough. He had them made at the Pic du Midi Observatory in the French Alps. They are so accurate- even down to magnetic contours extrapolated to the moon situation-that they can (and may) be used for our real moon landings.

Understandably, authenticity sometimes gets in the way of practical photography. All the expertise in the world couldn't solve the problems of filming an outer space movie in the gravitational field of earth. "Our biggest problem was weightlessness," says Masters. "How do you show people walking up walls and floating naturally in space?"

Usually, movie studios perform this kind of trickery with "matte shots"- two exposures superimposed. But Kubrick wasn't having any. Actions of people and the movement of floating objects would have looked phony. Hanging people on wires resulted in movements grotesquely unreal, and the wires glimmered.

Solutions were highly ingenious. In one sequence on the Discovery, one of the crew walks down a corridor, up a wall, across the ceiling and through a doorway upside down. To get it, the camera was bolted to the deck at the near end of the corridor. The actor, walking away from the camera to the far end, turned left and started up the wall. At that instant, the entire set rotated to the right taking the camera with it. The actor, of course, remained vertical as he walked around the revolving set, but on film he appears to be hiking up the bulkhead until he is standing upside down on the ceiling.

Actors learned to simulate the slow-motion actions of a zero gravity environment by studying Langley Research findings from moon gravity simulators. "In a real spaceship you'd use magnetic floors and shoes." says Ordway. "Walking movements would be slow and jerky. We discovered we could produce the same effect by walking people down an inclined plane. So that's how many of the scenes were shot. The camera was placed at the angle of plane to create the proper illusion." To heighten the effect, Velcro was used on floors and shoe soles so they'd grip each other like zippers.

There's no way to fake a spaceship centrifuge. So the giant revolving living quarters of Discovery, where most of the action takes place, was built full scale right in the studio. The 38-ton wheel, 42 feet across kissed the studio girders.

"We had to find a way to shoot film inside the centrifuge, showing it revolving," says Tony Masters. "The cameras had to show people walking every way including upside down."

The men devised revolving contacts for the cables so they wouldn't knot up. All services-including air conditioning were fed through the hub. Sixteen projectors inside the wheel provided readouts on the consoles. The camera rode on a remote-control dolly, its power cable dangling from the hub, allowing it to move counter to the wheel and stay out in front of the actors. Anchored to the rim of the wheel, it could take "upside-down" shots.

It wasn't all ice cream. "Lights around the centrifuge were big 10Ks and 5Ks," says Masters, "and they didn't like going around. They'd blow up inside the thing while we were shooting. Glass shattered and rained down on the actors."

For weeks Stanley Kubrick and Arthur Clarke fretted about whether in the airless lunar sky, you'd see far more stars than we do here. They finally decided that, since earth is 50 times brighter than the moon-viewed from the lunar vantage - it would wash out stars rather than make more of them visible, Surveyor proved them correct. "In fact," says Clarke, "Surveyor confirmed everything we had done."

How does this fantastic movie end'? You'll have to see it to find out. As this issue went to press, Kubrick and Clarke were still kicking it around. "The space explorers reach a moon of Saturn, all right," says Clarke. "As they are passing it, they discover what appears to be a king-size version of the magnetic monolith they unearthed on our own moon. As they approach, the great block performs an optical reverse and becomes, instead, a giant slot in the Saturnian moon's surface. As Discovery crosses the edge, the men see lights inside, but the lights turn out to be stars, and we take it from there to a hair-raising conclusion."

Courtesy of Popular Mechanics Originally Published in the April 1967 issue. Concept drawing by Lange & Clarke.

SATURN VEHICLE (below) has fission plasma engine. Crew in sphere is protected from nuclear unit (right) by 600-foot frame. Centrifuge in sphere provides gravity.

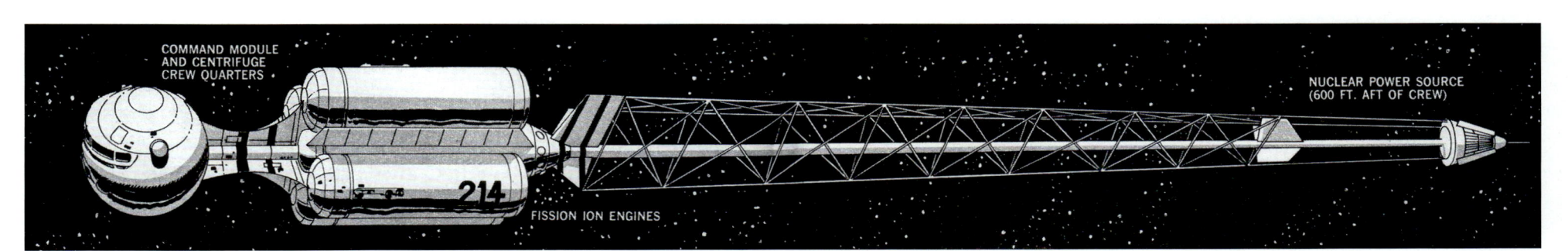

APPENDIX II

Technical information was provided by the following businesses:

Aero-jet General Corporation Covina, California
Instrumentation design and rationale, particularly for vehicle monitoring and display.

Aeronautical Chart and Information Center, St. Louis, Missouri and Arlington, Virginia
Charts of vast areas of the lunar surface, detailed data on Pic du Midi lunar photography, and support in obtaining such photography. Also, charts of the surface of Mars. In Arlington: photographs of Earth taken from high-altitude rockets and from satellites.

Air Force Cambridge Research Laboratories, Bedford, Massachusetts
Extreme altitude photography

Analytical Laboratories, Ltd., Corsham, Wiltshire, England,
Biological and medical instrumentation for centrifuge and for research panels for planetary and planetary moon probing.

Army Map Service, Washington, D.C.,
Maps of the Moon.

Barnes Engineering Co., Stamford, Connecticut,
Design concepts of telescopes and antennas, and their console instrumentation.

Bell Telephone Laboratories, Inc., Murray Hill, New Jersey,
Space Station V's picture- or vision-phone design, including rationale of routines to be followed in conducting orbit-Earth communications on a regular commercial basis. Assistance included typical jargon to be employed and suggested. Communications console for Discovery's centrifuge, including design and means of routine and non-routine transmitting and receiving.

Bendix Field Engineering Corp., Owing Mills, Maryland,
Control centers, consoles, and readout devices of manned space flight network.

Boeing Company, Aero-Space Division, Seattle, Washington,
Space simulation facilities information and photographs.

Chrysler Corp., New Orleans, Louisiana,
Interplanetary missions of scientific nature, particularly use of spaceship-mounted telescopes.

Computer Control Co., Framingham, Massachusetts,
Computer operations, terminology, console jargon.

Department of the Air Force, The Pentagon, Washington, D.C.,
Nuclear rocket propulsion.

Department of Defense, Washington, D.C.,
Color photography of Earth and general support in obtaining information of DoD space activities.

Douglas Aircraft Co., Santa Monica, California,
Instrumentation; vehicle design; console layouts; space vehicle films.

Elliott Automation, Ltd., Borehamwood, Hertfordshire, England,
Close support in supplying information on computer functions, readout materials, computer module design, computer terminology, and component miniaturization.

Eliot Noyes & Associates, New Canaan, Connecticut,
Cooperation in design and rationale as appointed agents of IBM in all computer sequences for Aries 1B and Orion, as well as spacesuit arm controls.

Flickinger, Don, M.D., Gen. USAF, ret., Washington, D.C.,
Appearance of the Earth from extreme altitudes.

Flight Research Center, National Aeronautics and Space Administration, Edwards, California,
Lunar landing research vehicle design and operation. Design and utilization of Gemini spacesuits.

General Atomic-Division, General Dynamics Corp., San Diego, California,
Propulsion system concepts for Discovery and capabilities of interplanetary spaceship.

General Dynamics-Convair, San Diego, California,
Films on Mars manned exploration missions; trajectory studies on manned interplanetary missions; mission mode concepts; advanced spacesuit design.

General Electric Co., Missile and Space Division, Philadelphia, Pennsylvania,
Design, instrumentation and rationale applicable to Space Station V, lunar roving vehicles, lunar bus design, instrumentation applicable to lunar base design, rationale, console instrumentation and operation of Discovery's propulsion; this is the system actually used. Detailed description of system and instrumentation.

George C. Marshall Spaceflight Center, National Aeronautics and Space Administration, Huntsville, Alabama,
Detailed photographic survey of the Marshall Center, including manufacturing and test areas; design and utilization of display and recording instrumentations; design of advanced space vehicles; design and implementation of NERVA engines, dozens of technical documents and photographs required during the film preparation.

Goddard Spaceflight Center, Greenbelt, Maryland,
General support surveying Goddard facility by photography; photo and information files on spacecraft, tracking systems, computers, instrumentation consoles.

Grumman Aircraft Engineering Corp., Bethpage, New York,
Apollo LEM mockup, detailed guided review, including instrumentation panels, Apollo missing planning; flight profile; activities of crew during entire mission; communications, etc., all applicable to Aries 1B and Orion.

Harbor General Hospital (A.T.K. Crockett, M.D., Chief of Urology), Torrance, California,
Consultation in the hibernation modules, monitoring devices. Note: Crockett is co-author of the paper "Total Body Hypothermia for Prolonged Space Travel." Ideas incorporated to extend, as modified by Ormand Mitchell.

Hawker Siddeley Dynamics, Ltd., Stevenage, Hertfordshire, England,
Provided the basic design of the interior of the space pods, including details of all panels. Several experts sent on three or four occasions and a number of meetings held at Stevenage with Clarke, Ordway, and Lange. Aided in antenna design and console instrumentation for antenna operation.

Honeywell, Inc., Minneapolis, Minnesota,
Assistance in instrumentation ranging from panels in Discovery and Pod Bay to the monitoring devices on the moon and the cockpit of Orion. Ideas were generated for the various docking sequences, leak detection aboard the Discovery, extravehicular activities, etc., etc. Also prepared a special report titled "A Prospectus for 2001 Interplanetary Flight." Also provided buttons and switch hardware.

Illinois Institute of Technology, Research Institute, Chicago, Illinois

International Business Machines, Armonk, New York,
Received very broad and valuable support from IBM through the making of the film, ranging from the design and construction by IBM subcontractors of computer panels and consoles, to the establishment of futuristic computer jargon and astronaut-computer interface. Also supplied valuable information on how computer-generated information would be displayed in the future. Hardware contributions: panels for the Aries 1B and the Orion cockpits plus buttons for many other sets, including Discovery's Command Module and centrifuge. IBM assigned Eliot Noyes & Associates, industrial designers (q.v.) as their consultants. Personnel made several visits to the M-G-M studios in London during the course of making the film.

International Business Machines, Ltd., London, England,
Provided direct technical input to the IBM-built panels; see IBM entry and that of Eliot Noyes & Associates. IBM U.K. Personnel visited the studios on a number of occasions, and many meetings were held in their offices. Detailed documentation on experiments that could be made from Discovery of the asteroids and the planet Jupiter and its twelve moons.

Institute for Advanced Study, School of Mathematics, Princeton, New Jersey,
Nuclear propulsion for Discovery.

Jet Propulsion Laboratory, California Institute of Technology, Pasadena, California,
Spacecraft information, photography of lunar surface mission, analysis of the asteroid belt and Jupiter fly-by probes.

Langley Research Center, Hampton, Virginia,
Detailed photographic tour of the center; gathering of large quantities of technical information relevant to *2001*, including photographs of laboratories, research vehicles, simulated docking and lunar landing devices, and film depicting appearance of man walking on the moon (simulator device). Considerable time spent in space station laboratory, viewing models and reports of space stations, and receiving briefings on rationale of space station technology.

Lear Siegler, Inc., Grand Rapids, Michigan,
Design concepts of advanced space vehicle instrumentation and display devices.

Libby, McNeil and Libby, Food Technology Research Center, Chicago, Illinois,
Food selections and menus for long space voyages; basis of menu selection for the centrifuge.

Lick Observatory, Mt. Hamilton, California,
Photography of the moon.

Ling-Temco-Vought, Inc., Dallas, Texas,
Reports on means and methods of displaying flight and other information to a crew undertaking interplanetary space mission.

Lowell Observatory, Flagstaff, Arizona,
Photography of the moon and planets.

Manned Spacecraft Center, National Aeronautics and Space Administration, Houston, Texas,
Detailed photographic survey of the center; reports and miscellaneous technical documentation on many aspects of manned space flight, with particular emphasis on Apollo lunar spaceship and space station technology. Very valuable cooperation in securing dozens of color photographs of the Earth taken from Mercury and Gemini spacecraft. Computer design and functioning; instrumentation; training techniques, astronaut routines, and conference room design and rationale utilized on lunar base sequence. Supplied six reels of Gemini tape in which mission control and pilot cross-talk was recorded. Maintenance and repair of space vehicles; Apollo mission rationale, time sequential analysis of crew activities and probable conversation with mission control, and advanced post-Apollo spacesuit design.

Martin Company (Now Lockheed/Martin), Baltimore, Maryland,
Technical instrumentation.

Minnesota Mining & Manufacturing Co., St. Paul, Minnesota,
A broad program of cooperation was outlined at original meetings in St. Paul.

Mt. Wilson & Palomar Observatories, California Institute of Technology, Pasadena, California,
Photography of the moon.

N.Y.U. College of Medicine, New York City, New York,
Development of techniques of placing man into hibernation and monitoring him when he is in the state. Very complete discussion of displays needed, design of the hibernaculums, a term devised by Dr. Ormand G. Mitchell, Assistant Professor of Anatomy, from whose many sketches were derived our final designs.

National Aeronautics and Space Administration Headquarters, Washington, D.C.,
Space station philosophy, effects of rotation on man; speed of rotation. Photography made by Ranger lunar probes; photography of space vehicles and NASA facilities; photography of planet Mars, general and overall support from NASA; capabilities of man as scientific observers during deep space voyage; continuing documentation of myriad subjects throughout progress of film.

National Aeronautics and Space Council, Washington, D.C.,
Feasibility of scene wherein a non-helmeted astronaut is very briefly exposed to space conditions.

National Institute of Medical Research, London, England,
Hibernation techniques and instrumentation.

North American Aviation, Inc., Space and Information Systems Division, Downey, California,
Photographs and documentation of the Apollo lunar spaceship, simulated lunar base experimentations; nature of the lunar surface.

Office of Naval Research, Brand Office, Embassy of the U.S.A., London, England,
Obtention of U.S. Navy full-pressure flight suit, including pressurization attachments, shoes, helmet; plus, technical documentation – all used in developing our own suits.

State of Oregon, Department of Geology & Mineral Information, Portland, Oregon,
Extraction of useful resources from lunar surface materials, utilizing SNAP nuclear reactors as heat source.

Philco Corp., Philadelphia, Pennsylvania,
NASA-manned Spacecraft Center Mission Control Center documentation, photography, and description of use of computer complex.

Royal Greenwich Observatory, Herstmonceaux, Surrey, England,
Design and rationale of the astronomical observatory and console in the centrifuge.

Societé de Prospection Electrique Schlumberger, Paris, France,
Geophysical instrumentation for the centrifuge. Cooperation included a meeting in Paris, two trips by personnel to London, submission of design concepts and rationale for use.

Smithsonian Astrophysical Observatory, Cambridge, Massachusetts,
Micrometeroid danger to space flight; means of detection; nature of space in terms of Discovery's flight through the asteroids.

Soviet Embassy, London, England,
Films of Soviet space programs. Stills of Luna 9 lunar photography.

U.S. Army Natick Laboratories, Natick, Massachusetts,
Data and photographs of space foods and associated equipment.

U.S. Naval Observatory, Flagstaff, Arizona,
Photography of the asteroids.

U.S. Weather Bureau, Washington, D.C.,
Detailed photographic coverage of the center; selection of documentation and photographs of appearance of Earth from satellite altitudes.

United Kingdom Atomic Energy Authority, Dorchester, England,
Instrumentation of nuclear reactor control consoles in the centrifuge and in the Command, Module. Meetings at Dragon reactor site and in studios.

University of Arizona, Lunar and Planetary Laboratory, Tucson, Arizona,
Photography and charts of the moon.

University of London, Mill Hill, Hertfordshire, England,
Advice on models of lunar surface; visit to studios and tour of laboratories at Mill Hill, including inspection of simulated lunar surface materials.

University of Manchester, Department of Astronomy, Manchester, England,
Photography of the moon from Pic du Midi sources; large scale photos of Tycho and Clavius craters; charts and maps of many areas of the moon; consultations on surface characteristics of the moon, nature of soil materials. Consultation on nature of celestial sphere as viewed from the Moon, i.e., the appearance of the heavens. Two meetings held in Manchester and one at the studios with members of Professor Kopal's staff.

University of Minnesota, School of Physics, Minneapolis, Minnesota,
Extreme altitude conditions, appearance of Earth from high-altitude balloons.

USAF School of Aerospace Medicine, San Antonio, Texas,
Photography of the Earth seen from extreme altitude manned balloons (Manhigh). Obtention of medical data in support of space medical aspects of film.

Vickers, Ltd. London, England
Engineered, constructed and assisted in the design of the rotating Centrifuge and monitoring equipment.

Vickers, Ltd., Medical Division, London, England,
Advice on hibernation and health-monitoring equipment and techniques for the centrifuge.

Whirlpool Corp., Systems Division (also known as RCA/whirlpool), St. Joseph, Michigan,
Development of the Aries 1B kitchen and planning of eating programs and routines.

Industrial Design products provided by the following:

Olivier Mourgue.
'Djinn Chairs' seen in the corridors of Space Station V. Built by Meubles Airborne S.A.

Geoffrey Harcourt.
'Modernist Club Chairs' seen in Clavius Moon Base meeting room.

Ernest Race.
'Heron Chairs' seen on the passenger deck of the Aries 1B. Re-upholstered in Mustard colored vinyl and leg support hidden by a cosmetic boundary between two sets of chairs.

Eero Saarinen.
'Tulip Table' seen in the corridors of Space Station V.

Arne Jacobsen.
Designed cutlery seen in *Discovery* eating area.

Seabrook Farms.
Packaging designs of food trays on *Aries 1B* and *Discovery*.

Hamilton watch company.
Modernist watches worn by pilots and astronauts.

Parker Pen Company
'Atomic Pens'. 4 were designed and served different functions, from burning to writing. Most advanced unit was the Robo-pen which could write in any language, and any style of penmanship.

Paris Match, Paris, France,
Supplied special futuristic cover for the magazine featured in Space Station V.

Eliot Noyes & Associates.
Under contract to IBM, designed layouts of the control panels on Orion III, Aries 1B and EVA pods. Designed the HAL9000 'glowing red eye', as well as the left arm-control pad on the space-suits.

P. Frankenstein & Sons (Manchester) Ltd. Aka Victoria Rubber Works.
An aerospace organization that constructed the spacesuits.

DuPont Chemical.
The space-suit material made of a metalized nylon fabric that is as pliable as silk.

ML Aviation Co. Ltd.
Supplied technical details and a prototype space helmet.

Mastermodels, London, England.
Constructed astronaut helmets, spacesuit front and back packs, the Rocketbus, and storage modules in the storage room of the podbay.

Boeing Aircraft Company
Assisted in the spheroid design, engine layout and landing gear of *Aries 1B*.

Grumman aerospace.
Assisted with outward appearance of EVA Pods.

Airfix model company, Revell USA model company, Renwal model company.
Supplied cases of model kits that were butchered and placed on the spacecraft miniatures to simulate functional surface details.

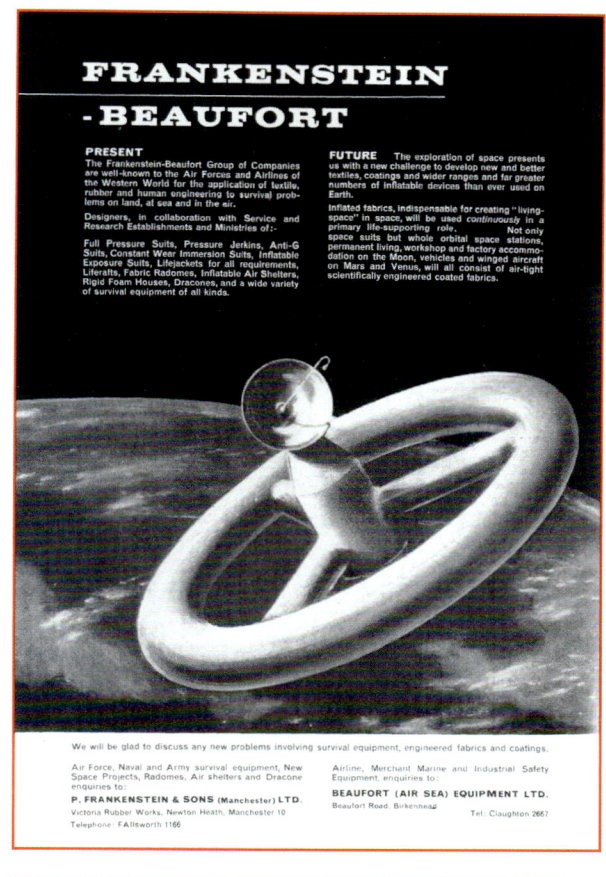

Ad from the early 1960s for the Frankenstein group who made the space suits for the movie.

Parker Pen company press ads. Parkers' unreleased press kits assuming they were going to market the various 'Atomic Pens'.

Whirlpool company press ads. Whirlpools' unreleased press kits assuming they were going to market the various 'Space Kitchens'.

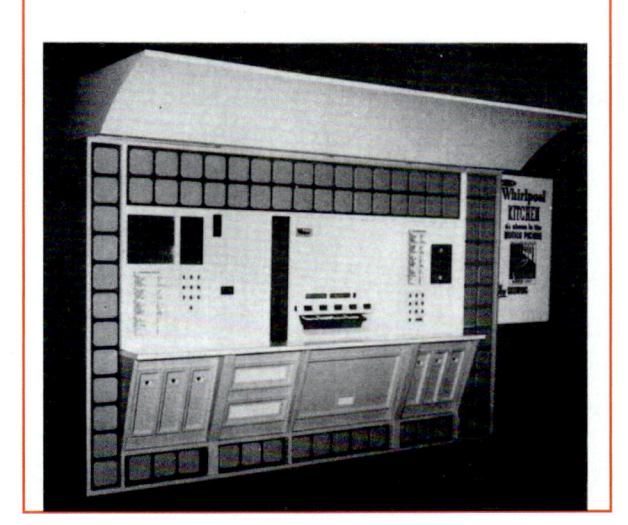

105

APPENDIX III

Honeywell – on the science & technology for '2001: A Space Odyssey'.

An abstract of the 100 page report: "A prospectus for 2001 Interplanetary Flight" prepared for Stanley Kubrick and Arthur C. Clarke by request of Frederick I. Ordway III.

By W.E. Drissel, Principal Systems Analyst, Systems and research Div., Honeywell, Inc. December 1967. Contributors: A. Macek, J. Miller & D. Stubbs. Artwork by E. Bermula. Edited by Adam K. Johnson for this publication.

Design engineers may dream a lot – but seldom with company encouragement. Yet this was the opportunity given to Honeywell's systems and research engineers, early in 1966.

The report was prepared to help assure the technical accuracy of a screenplay by Stanley Kubrick and Arthur C. Clarke on a movie entitled "2001: A Space Odyssey." This assistance was to take the form of developing ideas, working with an artist on drawings for a spacecraft of 35 years hence, and furnishing the technical language, dialogue and action pertinent to astronautics.

What made this assignment particularly intriguing was the fact that Honeywell was asked to assume that all technical impediments to its futuristic ideas would be overcome. With this in mind, engineers peered ahead to the year 2000 and came up with ideas for the following action sequences

- Test Procedures for the TMA-1

Designs and functioning of equipment that would be used to test an alien artefact– if actually discovered by humans. This would include chemical, physical, radioactive and atomic analysis by various specialized orthicons.

-Physiological Monitoring of Hibernation

Keeping humans alive over extended periods of time in deep cryogenic sleep.

-Spacecraft functions

Including: Rendezvous and docking procedures, manoeuvring, engine start-up and ignition, guidance and navigation, repair, leak detection, magnetic shielding, and probes that would be launched into Jupiter's atmosphere and moons.

-Computerised displays help astronauts in 2000

As a result of industry-wide efforts over the past 35 years toward integrated informational displays, pictorial and predictor displays now provide spacecraft pilots with information in immediately useful form. When the spacecraft power supply is functioning normally, the display gives periodic assurance of this fact; but should there be only marginal operation, or a failure, the display shows the faulty element as nearly isolated as possible and indicates with block diagrams or schematics how to repair it or select a back-up.

The spaceships' command center is dominated by a computer derived display complex which is controlled by voice or light pen and makes wide use of color. It is placed at arm's length from the pilots and nearly fills their field of view.

An automatic control phase, including plane change and gross catch-up, is

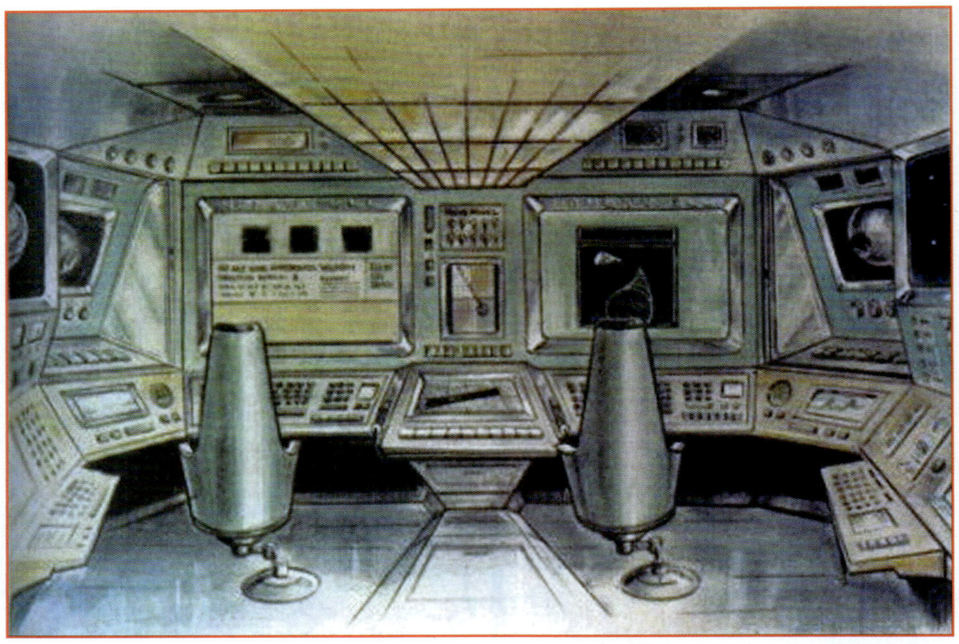

FIG 1 & 2 All spacecraft functions are monitored at the computer display complex. Subsidiary functions and possible activities such as recreation and training are also provided by the display. These designs would greatly influence the final design of Discovery's command deck. Deviating from dials, gauges and toggle switches was important to the futuristic concepts.

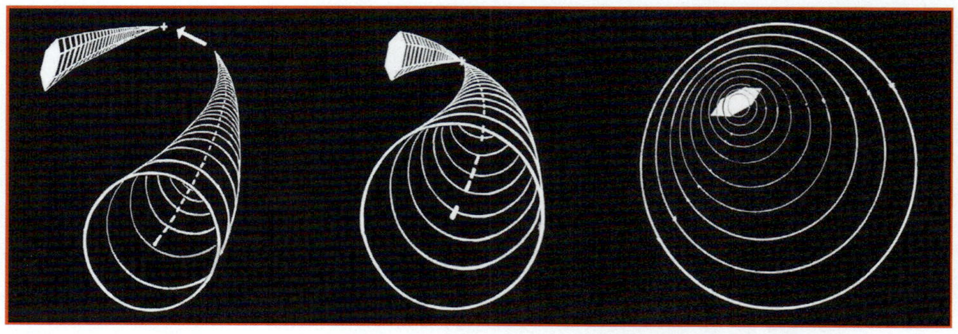

FIG 3 Manual rendezvous is accomplished through the use of a visual display. If any course correction is required during rendezvous, this is indicated by the error vector (arrow).

used for rendezvous with an orbiting space station. During this phase the display shows the position of the station and such things as how long before a velocity correction will be necessary, what the spaceship's attitude is, what the crew should be doing, what thrust is being developed, and how much fuel remains.

This phase is followed by a manual rendezvous phase using the spacecraft radar. At this stage the display shows the relative positions of the spacecraft and the space station, and if conditions are not correct for rendezvous, the pilots see a meaningful error vector; Fig 3 shows the display after the pilot has driven the error vector to zero.

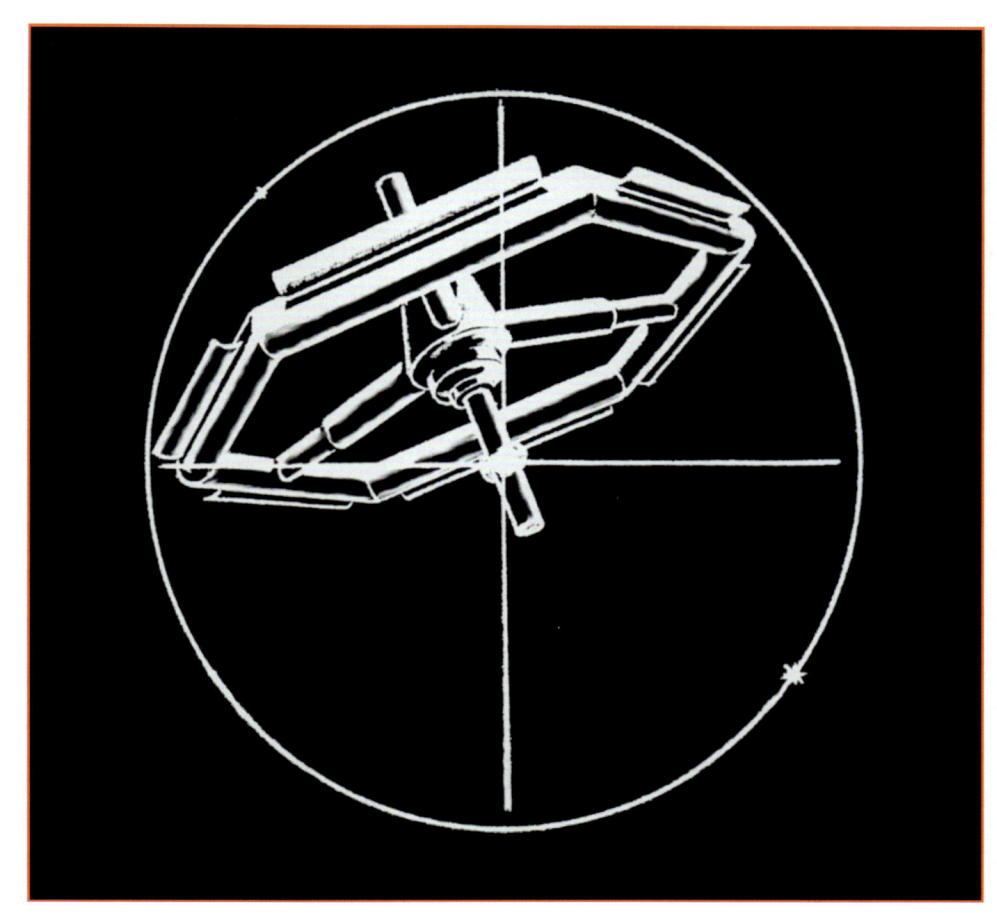

FIG 4 Docking is carried out manually by the astronauts. A high-resolution color TV picture of the target is displayed during docking. Error vector disappears when the spacecraft is placed on course for rendezvous. Display changes to a TV picture of the target when the spacecraft reaches a predetermined range.

This display continues until the target is within a predetermined range. Now the display switches to a high resolution color television picture of the target. Human beings are nowadays so good at docking that they usually handle it all themselves. The two small lights at the center of the docking display (Fig 4) are ranging lights. The cross hairs are positioned by the computer, so that the display follows the pilot's manoeuvres almost instantaneously.

-Events are displayed as checklist

Before every scheduled event, the pilots see an automatically displayed checklist. What the computer will do is listed in one category, the crew's tasks are in another. The computer performs a complete self-check first. This includes instructions to the pilots to assist it whenever necessary by voice or with the light pen. As each automatic check is completed, that item on the checklist turns green. The background of the checklist display remains suffused with pale green as long as all the tests are passed. Any tests which check out as marginal but not dangerously out of range, it shines red.

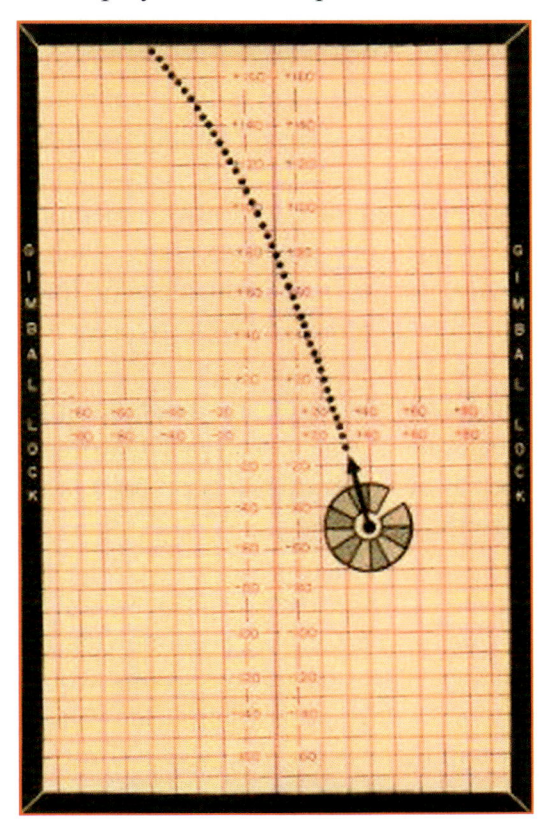

FIG 5 All attitudes are shown at once on the attitude display. The display uses a map of yaw and pitch projected on a plane instead of on a ball.

-Display indicates other conditions

When an out of tolerance condition occurs – for instance, when a velocity correction is required – and an appropriate section of the display panel changes to yellow, it indicates to one of the pilots that his participation is required to correct the condition. Information on what corrective action he must take is displayed at the same time. With this guidance, the crew can easily modify the out-of-tolerance condition, again either by voice or with a light pen. The voice controllers which convert the crew's oral commands into electrical inputs to the computer have a speech-recognition capability sufficient for any commands that might be given.

If there is a countdown prior to any of the preplanned mission sequences, the countdown is also displayed along with any simultaneous, automatic checks.

-Cutting edge attitude display used

Apart from the main display panel and the information it shows, the pilots' main object of interest is the attitude indicator. It indicates the attitude of the spacecraft with respect to a set of reference axes, with the map of yaw and pitch projected on a plane instead of a ball.

This technique was the outcome of simulation studies at Honeywell many years ago. They showed that test pilots had trouble manoeuvring from a position on the "front" side of the ball to one at the "back." The pilots found difficulty in determining which way to start a manoeuvre or how to apply corrections once started. Therefore, an engineer conceived a display (Fig 5) that shows "all attitudes at once," in this manner:

Lines of constant pitch are shown horizontally and lines of constant yaw vertically;

The opening on the ring shows roll attitude;

The location of the ring's center shows the pitch and yaw attitude; and

The dashed line depicts the predicted changes in attitude which would result from present attitude rates.

During limit cycling, the dashed line waves back and forth as the spacecraft reaches the attitude limits and the control jets pulse. If the flight requires a particular attitude, a light spot appears at the desired attitude. If manual control of attitude is required, the arrow gives the pilot an indication of the direction of required acceleration to accomplish the needed change. If automatic control is required, the fuel optimal-attitude control can be computed in advance and executed on command. The display then shows the predicted control path so that the pilot can monitor the manoeuvre.

-Recreational and training displays included

The tedium inherent in long interplanetary flights is combated with a variety of recreation. The computer, for example, is able to store each crew member's favorite books, new books on his favorite subjects and his favorite games.

Machine played games are of such sophisticated design that they can challenge each man at his level of ability and offer him a predetermined chance of winning. Automated bridge games for 1, 2, 3 or 4 players are available on the spaceship. The machine deals the cards and keeps track of play. For convenience the dummy is read from a scope rather than using magnets or other contraptions to prevent the weightless cards from floating all over the cabin. The machine keeps score and fills in for any missing players. Similar arrangements enable the crew to play chess, checkers, poker, gin rummy and a number of other games.

-Extravehicular work without a spacesuit

For work outside the spacecraft, the astronaut gets into a manoeuvring unit like that shown in Fig 6. The two wing-like structures on the unit permit a nearly complete range of natural arm motions.

Inside the unit, the astronaut is in a shirt sleeve environment. This is felt to be necessary for any significant amount of extravehicular work because the strength that he requires to flex a pressure suit represents a large fraction of the total force that he can deliver over an extended length of time.

A complex of bioforce sensors is attached to the astronauts' arms and hands. The forces generated by the manipulators are fed back against these force sensors to give him the sensation of feeling. He can vary the gain of each loop as he works, to minimize the work load or maximise his "feel." Since his hands are occupied by the force sensors, he controls his attitude and translational position by voice. The voice controller also is the means for changing the gains of the bioforce sensors. Or, for example, the astronaut feels that the work requires extra grip forces from his right hand, he commands: "Right grip times two," and the voice controller doubles the gain of the right-grip bioforce sensors.

The structure located on the midline of the manoeuvring unit is a device for keeping the unit in position at the work site. Past studies showed that it consumes too much fuel to use thrusters to counteract tool forces. The spacecraft, therefore, has exterior attachment points from which every point on the hull can be reached.

-What will all this require?

Back to 1967 and contemporary reality. What is needed to make the foregoing prophecies come true? The display that was dreamed up uses more saturated colors than can be found in the gamut of today's three-color systems; a four to six-color system would be necessary. If the display were a six-color oscilloscope, considerable improvement over today's capability would be required in alignment and deposition of phosphors. The panel might perhaps be a flat electroluminescent panel. If so, a fine-grained display structure, with addressable elements is required. The data rate to drive such a display far exceeds the present capacity of computer communication channels. Even if the display unit contained its own memory and communication, present-day computers could not load the memory fast enough.

The computer is in many respects merely an extension of today's computers albeit with an enormous memory and an increase in speed. Perhaps the only way at present to visualise a display capable of meeting all the requirements described is to imagine several autonomous computers with shared memory. Maybe by 2000, there will be a self-organising multiprocessor. Such a device would decide when more or fewer processors were needed "on line" and then switch instruction registers, arithmetic units, and so forth on or off.

The 2000 computer would necessitate a very elaborate signal-conditioning interface to perform all the tests of equipment states. The fact that many different functions would be performed at once indicates a need for a large versatile interrupt and input/output system.

Finally, the spacecraft computer, like the attaché case computer, would be under "systems" control at all times. For this reason, the requirements for software would be extremely demanding. The software programs would be exceedingly large, with a multitude of very important but seldom used emergency loops. Perhaps by 2000, it will be possible to generate software directly from an English-language requirements document.

The astronauts manoeuvring unit would require sensitive, accurate and light bioforce sensors, which would have to be easy to doff and don. Investigations currently in programs may lead to the breakthrough.

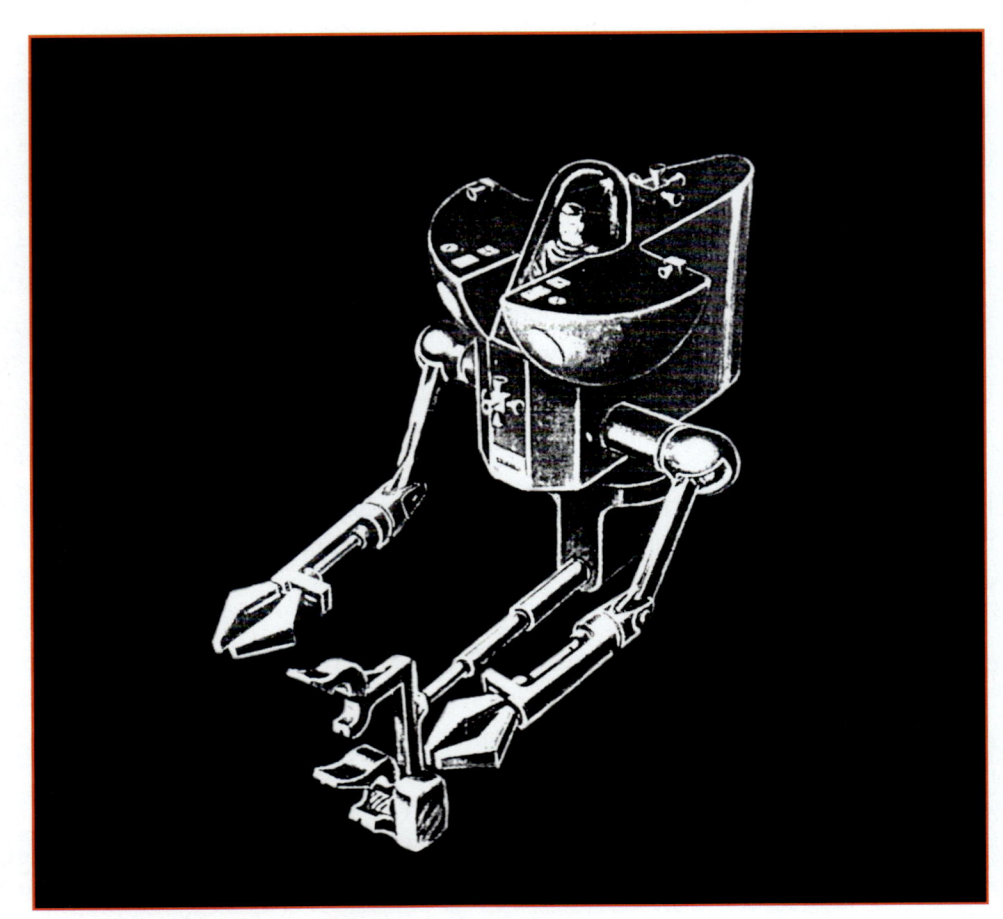

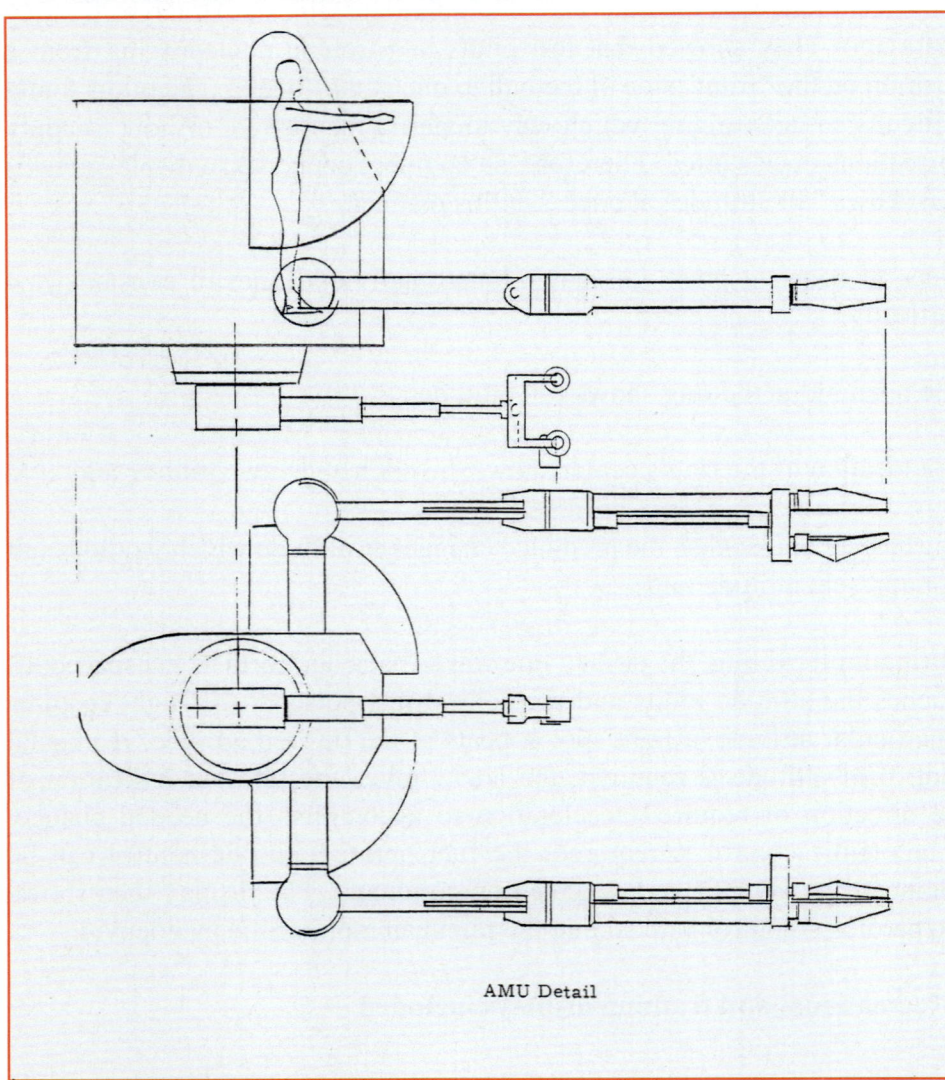

FIG 6a & b An astronaut uses his voice to control the attitude and translated position of the manoeuvring unit. His hands are occupied by force sensors which give him the sensation of feel as he operates the manipulators.

Long space flights are also an opportunity for intensive study. A 1-2-year flight is ample time to prepare for a master's or doctor's degree. Furthermore, the complexity of the spacecraft means that the crew must constantly review procedures, schematics and instructional texts. Manual skills like spacecraft manoeuvring, emergency operations, donning and doffing of pressure suits and extravehicular maintenance, have to be practiced periodically. The computer and display keep tabs on these practice sessions and make comparisons with past performance norms.

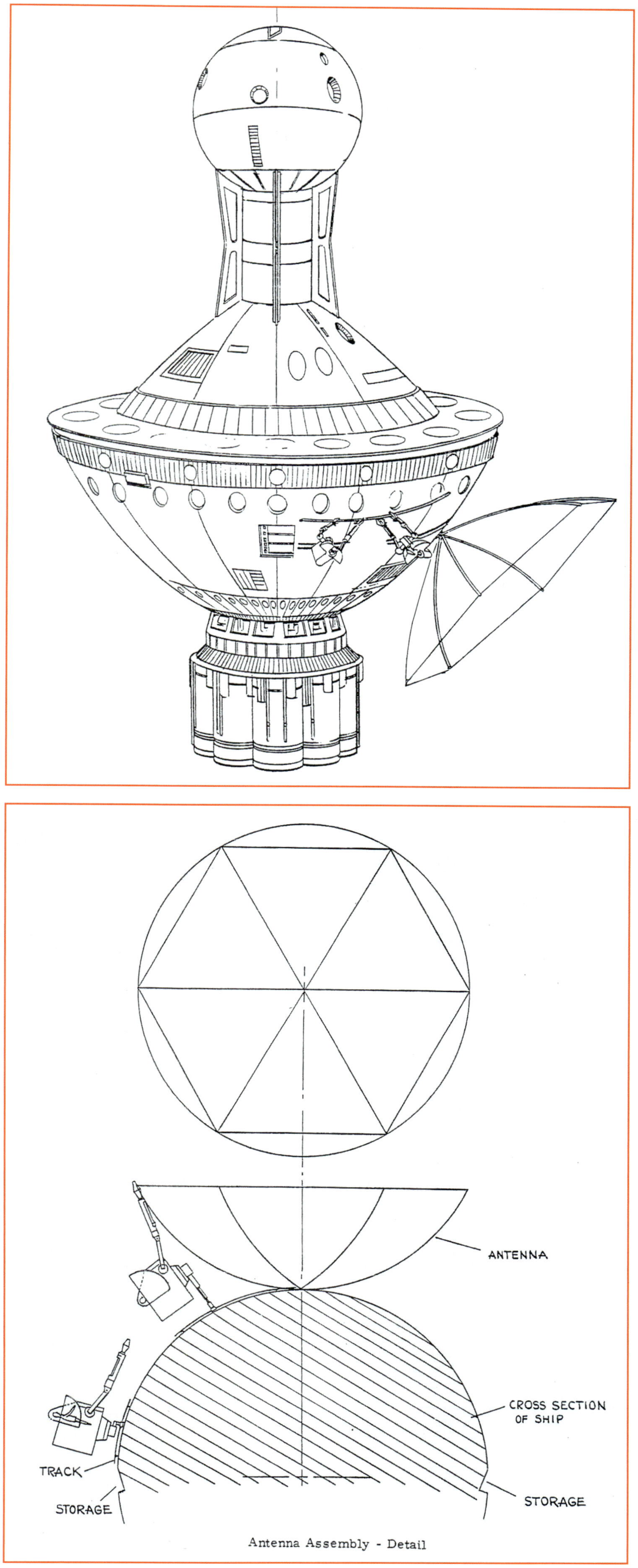

The manoeuvring unit is equipped with a voice controller. Studies at Honeywell have indicated that this is the best way to control a unit's attitude and position, because the astronaut's hands would be occupied. Some devices already exist today that would serve as a voice controller if the vocabulary were small enough. Power requirements are high for some of these devices though, and all need to have their size reduced and recognition-accuracy and versatility improved.

All these technical advances are relatively simple compared with the task of building the attaché case. This involves a display, telecommunications system, computer with interchangeable storage, and a line printer – all compressed into 0.75 cubic feet, weighing less than 25 pounds and self-powered.

FIG 8 Probes launched from the lower sphere of Discovery perform a great variety of tasks. Atmospheric Probes: Launched into Jupiter's atmosphere contain an optical TV and infrared camera, charged particle detectors, magnetometers, accelerometer, microwave detectors, sonic anemometer, and radar altimeter. Soft Landing Probes: Launched onto an 'atmosphereless' moon contain a gravity sensor, optical TV and infrared camera, X-ray detector, charged particle detectors, magnetometers, accelerometer, microwave detectors, electron probes, sonar receiver, radar altimeter and a micrometeorite detector.

Electronics galore in attaché case

The space-travelling executive in 2000 has available a revolutionary attaché case, though because of its high cost it is used by few people. Its principal features are:

- Display
- Telecommunications system
- Computer
- Microstorage files
- Control keyboard and line printer

The display is computer-derived in color and viewable in daylight. The display carries TV pictures derived from the case's telecommunications system, as well as the context of the microstorage file. As the owner dictates notes to be entered into the file or typed out in letter format, the display shows the words spoken so far. Editing is accomplished with a light pen, by voice or through keyboard instructions.

The telecommunications system includes a telephone handset and dial for voice communication, a TV camera, an automatic "linefinder" which scans the electromagnetic frequencies during transmission, a scrambler for message security, and a link to the computer to provide automatic transmission of large blocks of data. The TV camera can operate either cabled to the attaché case or through a microwave or laser link.

FIG 7a & b Two AMU's erect the master communications antenna on the space ship (early concept design of Discovery as envisioned by Honeywell).

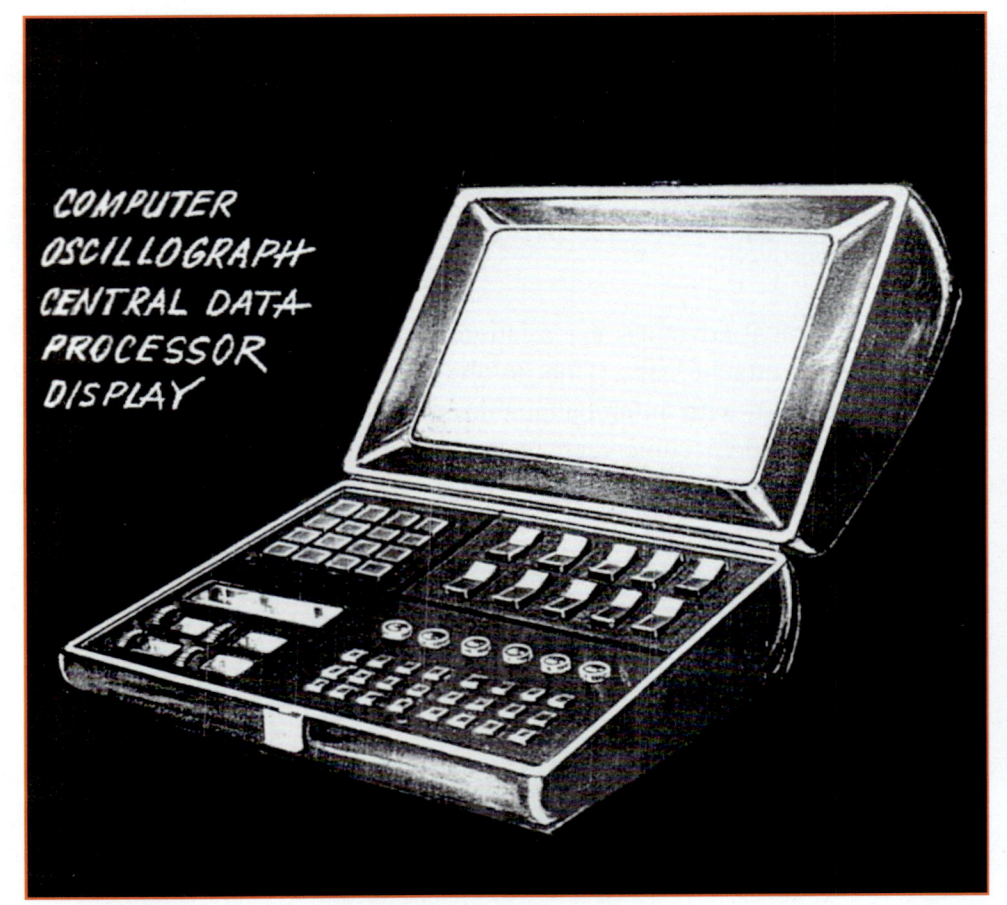

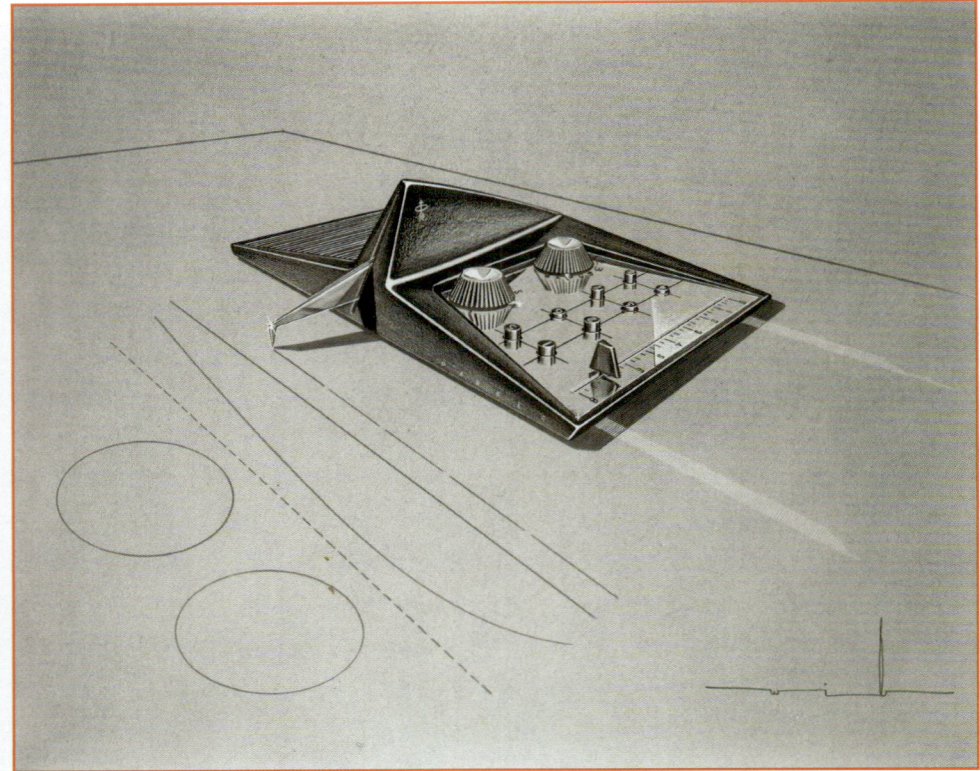

Artist's rendering of Parker Robo-Pen designed for the film.

The computer employs a highly developed system program, for it is under system control at all times. There are no source or object programs from outside. Because the computer is primarily concerned with data handling, both it and its programs lean heavily toward list processors, high-speed search routines and editing programs.

The microstorage files are very high-density, random-access units that can easily be plugged in and unplugged.

The keyboard, light pen and handset microphone allow user and computer to communicate. The lineprinter is a high speed photographic unit driven by the computer.

FIG 9 Attaché Case is crammed with electronic gadgets to make the executive a true space age traveler. Despite its conventional size, the case contains a color display, a telecommunications system, a computer with interchangeable storage units and a line printer. The user can communicate with the computer by means of the printer keyboard, a light pen, or a handset microphone.

REFERENCES

Further reading/viewing:

2001: A Space Odyssey. A film by Stanley Kubrick . Warner Home Video, Blu-ray disc (2007)

2001: Space Odyssey. A novel by Arthur C. Clarke. (1968)

The Lost Worlds of 2001. By Arthur C. Clarke. (1972)

2001: Filming the Future. By Piers Bizony. (2001).

The Stanley Kubrick Archives. By Alison Castle. (2008).

Cinefex (magazine) issue # 85 "2001: A Time Capsule" by Don Shay and Jody Duncan. (2001)

Bibliography:

The US Space & Rocket Center historical archives, Huntsville, Al.

Spaceflight magazine. (March 1970).

History of Rocketry and Space Travel. By Wernher von Braun, Frederick I. Ordway III, Harry Lange. (1966).

American Cinematographer Vol 49, no6. Herb Lightman. (1968).

Missions and Flight Operations for Nuclear Rockets, Report LMSC-A653019 Lockheed Space Division (Nov. 1964)

NERVA engine performance and design reports from the Marshall Space Flight Center, Huntsville Al. (1962 – 1966).

IBM's early computers. Bashe, Johnson, Palmer, Pugh. (1985).

Introduction to Artificial Intelligence: Second Edition (1985).

What computers still can't do: A critique of artificial reason. Hubert L. Dreyfus (1992).

Optical Memory Storage and retrieval. US patent# 3765749.

'2001: A Space Odyssey' by Frederick I. Ordway III, AAS publications (1985)

'A prospectus for 2001 Interplanetary flight' Honeywell aerospace division (1966)

'Intelligence in the Universe' by Frederick I. Ordway III and Harry K Lange (1966)

ACKNOWLEDGEMENTS

Dr. Frederick I. Ordway III, Science advisor on the film '2001: A Space Odyssey'.

The Staff at the US Space & Rocket Center Historical Archives.

Mr. Frederick Barr

Mr. Andrew Birkin

Mr. Robert Godwin

Dr. David Thornhill

Mr. Jeff Wargo

Mr. Tony Hardy

Ms. Irene Willhite

The British Interplanetary Society

Pan Am is a registered trademark of *Pan Am Systems*

Aeroflot is a registered trademark of *Aeroflot Airlines*

Cover design by Frederick Barr. Insert pictures by Adam K. Johnson.

Editor: Robert Godwin

Photo Credits:

Stanley Kubrick & Geoffrey Unsworth

Harry Lange

John Jay, Keith Hamshere, Ken Bray

Dmitri Kessell

Frederick I Ordway III

Warner Brothers Pictures

Andrew Birkin

Original MGM drawing (Draughtsmen) credits:

John Graysmark, Alan Tompkins, John Siddall, Tony Reading, Frank Wilson, Peter Childs, Wallis Smith, Martin Atkinson, Brian Ackland-Snow, Alan Fraiser, Peter Jarrut, John Fennar.

Original MGM drawing Digital Restorations
© 2012 Adam K. Johnson.

3-D Centrifuge graphic ©2012 Robert Godwin

ALL RIGHTS RESERVED. This book contains material protected under International and Federal Copyright Laws and Treaties. Any unauthorized reprint or use of this material is prohibited. No part of this book may be reproduced or transmitted in any form or by any means, electronic or mechanical, including photocopying, recording, or by any information storage and retrieval system without express written permission from the author & publisher.

All images in this book come from private collections that are owned by the US Space & Rocket Center historical archives for preservation. A thorough copyright search was executed by the author before publishing. No copyright infringement is intended or implied.

Extracts from the Journal of the British Interplanetary Society, Honeywell and the Hearst Corporation have been used with permission.

Please respect the *Digital Millennium Copyright Act* regarding the content of this book.

A portion of the sales of this book goes to the *US Space and Rocket Center* to help maintain these important historical archives.

2001: The Lost Science by Adam Johnson
©2012 Apogee Prime/Adam Johnson

ISBN 978-1926837-19-2
First Edition
Printed in Canada

Published by Apogee Prime a Division of Griffin Media

To purchase copies of the accompanying movie "2001: The Science of Futures Past" please visit apogeeprime.com

"2001: The Science of Futures Past"
**an exclusive new documentary by award-winning film maker
Michael Lennick of Foolish Earthling Productions.**

Produced, written and directed by Michael Lennick

Narrated by Jack Senett

Original music by John Herberman, Yuri Sazonoff and Johanne Strauss II

Animators: Uchenna Nwaehie, Mike Riley and Robert Godwin

Cinematographer/editor: Michael Lennick

Transcripts by Makala Kalyn

Executive Producers: Michael Lennick and Robert Godwin